
SELECTED READINGS IN PHYSICS
General Editor: D. TER HAAR

HELIUM 4

HELIUM 4

BY

ZYGMUNT M. GALASIEWICZ

Professor of Theoretical Physics, University of Wrocław,
Institute for Low Temperatures and Structure Research
of Polish Academy

PERGAMON PRESS

OXFORD · NEW YORK · TORONTO
SYDNEY · BRAUNSCHWEIG

Pergamon Press Ltd., Headington Hill Hall, Oxford
Pergamon Press Inc., Maxwell House, Fairview Park, Elmsford, New York 10523
Pergamon of Canada Ltd., 207 Queen's Quay West, Toronto 1
Pergamon Press (Aust.) Pty. Ltd., 19a Boundary Street, Rushcutters Bay, N.S.W. 2011, Australia
Vieweg & Sohn GmbH, Burgplatz 1, Braunschweig

First edition 1971
Library of Congress Catalog Card No. 71-128037

Printed in Hungary

08 015816 1

CONTENTS

Theoretical Papers

PREFACE

HELIUM 4 is an especially interesting element because of the amazing property of superfluidity. This phenomenon was discovered in 1938 and a satisfactory theory was given in 1941. Recently we have seen the fruitful development of a microscopic theory of superfluidity. The aim of the Introduction is to pay as much attention as possible to this development.

The limited size of this book has meant many restrictions in the list of selected papers. Still, it is felt that few important papers have been omitted.

This book was written during a stay at Copenhagen and I should like to express my gratitude to Professor Aage Bohr for special hospitality. I would like to thank the authors and publishers for permission to reprint or translate the original articles. Finally, I would like to thank my wife who helped me with the preparation of the book.

Copenhagen ZYGMUNT M. GALASIEWICZ

PART 1

INTRODUCTION

HELIUM 4 was discovered in 1868 in the Sun's atmosphere. The yellow helium D_3 line was seen during an eclipse by many observers. Only one year later, in 1869, Rayet pronounced that the gas emitting the D_3 line was a new element like, for example, hydrogen.

The French Academy considered the merits of the helium discovery and ordered a medal with the effigies of J. Jenssen and N. Lockyer. The first mention of the name "helium" can be found in the literature in a footnote to a paper by Kelvin (1871) published in *Nature*: "... Frankeland and Lockyer find the yellow prominescences to give a very decided bright line not far from D, but hitherto not identified with any terrestral flame. It seems to indicate a new substance, which they propose to call Helium."

Finally, only in 1895, helium was discovered on Earth. It was done by Ramsay who obtained it from the mineral cleveite. Helium 4 occurs also in the atmosphere with a concentration of about 1 part in 200,000.

A very important property of helium is that the attractive van der Waals forces are weaker than in all other substances including hydrogen. The van der Waals forces try to bind the molecules in liquids and solids. In solids these forces lead to the spatial ordering manifested in the formation of a crystal lattice. Kinetic (thermal) energy of the molecules acts in the opposite direction, i.e. it tries to destroy the ordering. Solidification is a result of the balance between the potential energy connected with the van der Waals forces, and the kinetic energy of the molecules. From the classical point of view, at absolute zero, the kinetic energy of particles is equal to zero. The potential energy must have a minimum which is realized for a crystal lattice of suitable shape. From the quantum mechanical point of view, however, the kinetic energy of a system does not vanish at absolute zero. It is called "zero point" energy and describes the so-called zero-point motion. Usually the zero-point energy is small compared with the potential energy. But just for helium and hydrogen it is large in comparison to the potential energy arising from the van der Waals forces. London (1936, 1939) has calculated the zero-point energy for hydrogen and helium. This energy is a decreasing function of molar volume, favouring a large molar volume. The potential energy can also be expressed as a function of molar volume. The form of potential energy for liquid helium London took from papers by Keesom and Taconis (1937, 1938) and for solid helium from Yntema and Schneider (1950). In order to calculate and compare the total energies London assumed that the zero-point energy is the same for the solid and liquid state. He obtained the very important result that at absolute zero the total energy for liquid helium lies lower than for solid helium. The minimum of the total energy is realized for a molecular volume of about 28 cm^3 which agrees with experimental data. (Recently the most detailed calculations of the energy are given by McMillan (1965)). The same considerations applied to hydrogen give a total energy which is smaller for the solid than for the liquid state. The reason is that the van der Waals forces in hydrogen, although small, are about twelve times larger than for helium.

Helium cannot be solidified under its own pressure by cooling only. When approaching absolute zero it can be solidified at a pressure of about 25 atm. In the solid state helium exists in three allotropic crystal structures: hexagonal close packed (h.c.p.), face-centered cubic, and body-centered cubic. Near absolute zero the X-ray diffraction experiments show that the structure of solid helium is h.c.p. This structure as a possibility for helium also under normal pressure was considered by London in his calculations, because he was using the Yntema–Schneider (1950) potential.

From the above considerations we see that at absolute zero, without applying higher pressure, helium remains liquid. This fact is due to the zero-point energy which is of quantum mechanical origin. Therefore we can call liquid helium a quantum liquid and in our investigations expect a manifestation of new, special properties.

The most interesting properties of helium 4 are observed after liquefaction and then cooling to temperatures nearer absolute zero. The liquefaction of helium itself was extremely difficult and helium was the last of the gases which one succeeded to liquefy.

The greatest success in liquefaction of gases were the achievements of Faraday in 1845 but the first hint about liquefaction was given already by the famous satirist Lucian of Samosata (second century A.D.): "... I write here of events that did not happen, that I never witnessed and that I did not learn from others. I speak of things that never had nor could have any substance, and I urge my readers not to give them any credence.... Now I must tell you the novel and extraordinary things I noticed during my stay on the moon.... Their drink is air compressed into a vessel, and out of which they squeeze a kind of moisture like dew...."*

However, only about 2000 years later, in 1883, air (oxygen, boiling temperature 90·1°K, and nitrogen, boiling temperature 77·3°K) was liquefied by Wroblewski and Olszewski. It was at last the breaking of the impasse which lasted from the Faraday's times, i.e. for 38 years. During this time the physicists shared the opinion of Lucian that air cannot be liquefied. The same was assumed about hydrogen and helium. Hydrogen was condensed finally in 1898 by Dewar (boiling temperature 20·38°K) and helium in 1908 by Kamerlingh Onnes (boiling temperature 5·22°K). The last date is the birthday of low-temperature physics.

By measuring the density of liquid helium as a function of tem-

* The author is very thankful to Professor L. Rosenfeld for translating this fragment into English.

perature Kamerlingh Onnes (1911)* found the strange property of helium that at about 2·2°K there seems to be a maximum in its density. It means that by lowering the temperature from 2·2°K to absolute zero helium starts to expand. The more accurate measurements performed by Kamerlingh Onnes and Boks (1924, reprinted in Part 2 as paper E1 on p. 59) confirmed the existence of the density maximum in the neighbourhood of 2·3°K. The most important conclusion was that the density–temperature curve has a sharp discontinuity at the maximum.

Three years later Keesom and Wolfke (1927, reprinted in Part 2 as paper E2 on p. 65) observed that also the dielectric constant shows a sudden jump in the vicinity of 2·3°K. "...The thought suggested itself that at that temperature liquid helium transforms into another phase, liquid as well. If we call the liquid stable at the higher temperatures 'liquid helium I', the liquid stable at the lower temperatures 'liquid helium II', then the dielectric constant of liquid helium I should be greater than that of liquid helium II...."

Further arguments for the existence of two modifications of helium were given first in a paper by Dana and Kamerlingh Onnes (1926), describing the observed jump in the heat of vaporization, and second in a paper by Van Urk, Keesom and Kamerlingh Onnes (1926) which observed a jump in the value of the surface tension. The next discontinuity, which is of the utmost importance, was noticed by Keesom and Clusius (1932), reprinted in Part 2 as paper E3, p. 72. They measured the temperature dependence of the specific heat. A jump was observed at the vicinity of 2·19°K. The shape of the specific heat versus temperature curve resembles the

* It is interesting to notice in this connection that the paper of Kamerlingh Onnes (1911) entitled "Further experiments with liquid helium" had a second part "On the change in the resistance of pure metals at very low temperatures, etc. III. The resistance of platinum at helium temperatures", which was an important step towards the discovery of superconductivity which occurred in the same year.

shape of the letter λ and therefore the singular point at this curve is called λ-point. It was Ehrenfest who suggested this name. At the λ-point no latent heat, characteristic of a first-order phase transition was observed.

The existence of the λ-point and the fact that liquid helium remains liquid at absolute zero until a pressure of about 25 atm is

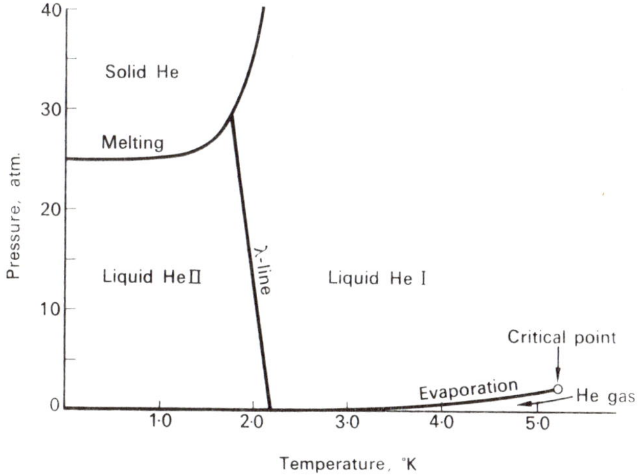

FIG. 1. Diagram for ^{4}He in the p-θ-plane (after London, 1954).

applied leads to the special form of the diagram for helium in the $p-\theta$-plane.

We see that the ordinary triple point between the solid, liquid and gaseous phases does not exist for helium. But there exist two other triple points at the ends of the λ-line which separates liquid helium I from liquid helium II.

The results of the paper of Keesom and Clusius (1932) led to an attempt to extend the then existing ideas about phase transitions. This was done by Ehrenfest (1933). He introduced the idea of an nth order transition when the Gibbs potential $G(p, \theta)$ has a discontinuity of its nth order partial derivatives. This discontinuity must occur at the transition line $p = p(\theta)$. Ordinary phase transitions are usually first order. The transition from helium I to helium II at the λ-point, that in superconductors at the critical temperature, and that in ferromagnetics at the Curie temperature are second-order transitions. The transition occurring at the Bose condensation is a third-order one.

From the above description of the group of experiments it is clear in all aspects that at the λ-point we have to be concerned with a second-order phase transition.

The next group of experiments concerned the properties of helium II called by London (1954) "super" properties.

The first one was discovered by Keesom and Miss Keesom (1936, reprinted in Part 2 as paper E4 in p. 89). They found a very large thermal conductivity of helium II. Its maximum is attained at 1·9° and equal to 810 cal cm^{-1} sec^{-1} $(°\text{K})^{-1}$. It is, for example, 10^3 times larger than the thermal conductivity of pure copper. Such a large conductivity was called by the authors "super heat conductivity" and by London (1954) "thermal superconductivity". One year later Allen, Peierls and Uddin (1937, reprinted in Part 2 as paper E5 on p. 92) noticed that the heat conductivity in helium II is quite different from that in other fluids. New experiments studying the heat flow in helium II were performed by Allen and Jones (1938, reprinted in Part 2 as paper E6 on p. 95). They discovered the so-called fountain effect. In the experiment an open bulb with a capillary in the bottom part was immersed into the helium bath. If the heat was introduced electrically by means of the coil inside the bulb the level of the liquid rose in the bulb. In this experiment the increase of the temperature produces the pressure difference between bulb and reservoir.

In the second variation of this experiment a glass tube was used widened at the helium reservoir side and packed with fine emery powder. If the powder was heated by radiation one observed a fountain of liquid emerging from the tube.

The inverse of the fountain or thermomechanical effect is the mechanocaloric one discovered by Daunt and Mendelssohn (1939a). A small Dewar with a hole at the bottom end was filled with emery powder. If a pressure difference was produced by withdrawing the Dewar from the helium bath, the helium flows out of the Dewar and a resistance thermometer indicated an increase of temperature in the Dewar. When lowering the Dewar into the bath, helium flows into it and a decrease of temperature was observed.

The measurements of the viscosity of helium II performed by Kapitza (1938, reprinted in Part 2 as paper E7 on p. 99), by using some modification of the capillary flow (narrow channels) method show that it is vanishingly small. The upper limit estimated from these experiments shows that the viscosity of helium II must be 10^4 times smaller than that of hydrogen gas. "... The present limit is perhaps sufficient to suggest by analogy to superconductors, that the helium below the λ-point enters a special state which may be called a 'superfluid'...."

The viscosity of helium II was also measured using an ordinary flow method by Allen and Misener (1938). They also found the vanishingly small viscosity. Furthermore, they observed that the flow velocity is practically independent of the pressure difference at the ends of capillary. "... The observed type of flow, however, in which the velocity becomes almost independent of pressure most certainly cannot be treated as laminar or even as ordinary turbulent flow...."

Notice that in laminar flow the average velocity is proportional to the pressure difference and in turbulent flow proportional to the square root of pressure difference. Thus, in the light of Allen and Misener's experiments the superfluid flow seemed to be more closely related to turbulent than to laminar flow.

The results of Kapitza (1938) and Allen and Misener (1938) seemed to be in contradiction to those obtained by Keesom and MacWood (1938, reprinted in Part 2 as paper E8 on p. 103). They found that below the λ-point the viscosity of helium does not differ significantly from the viscosity of helium I. From the point of view of the ordinary liquid it made no difference that the rotating discs method used in the last experiment was quite different from the methods used earlier.

The last of the "super" properties of helium II is the existence of what London calls "supersurface films". The existence of such films covering all walls in contact with helium II was ascertained by Daunt and Mendelssohn (1938a, reprinted in Part 2 as paper E9 on p. 111) and independently by Kikoin and Lasarev (1938). The last authors found a film thickness of about 10^{-5} cm. The best measurements so far are those performed by Ham and Jackson (1954).

The existence of the films was suggested by Rollin and Simon in 1936. We quote from a paper by Kürti, Rollin and Simon (1936): "... We have also found that the vessels containing liquid helium cooled below the λ-point and connected by a tube to the warmer parts of the apparatus show an abnormal high heat inflow. This behavior can probably be explained by the existence of a liquid layer along the walls of the tube..."; and from a paper by Rollin and Simon (1939): "... It is concluded from the experiments that liquid helium II 'creeps out' of the containers in the form of a film and that this is responsible for the anomalous effect observed.*..."

The existence of the films can explain also the result of an older effect observed by Kamerlingh Onnes (1922). If a small Dewar with liquid helium was immersed into a greater vessel with liquid helium there took place a transfer of liquid until the levels in the two vessels were the same.

Thus many surprising and interesting properties of helium discovered experimentally and gathered for years waited for a general theoretical explanation. Fröhlich (1937) was the first to suggest that the λ-transition is a special kind of order–disorder transition. The model of the transition proposed by Fröhlich was criticized by London (1938a); he considered the Bose–Einstein condensation

* Anomalous evaporation from vessels containing liquid helium.

as a peculiar example of the order–disorder transition. He paid attention to the curve describing the temperature dependence of specific heat for an ideal Bose gas. This curve has at some critical temperature a discontinuity. It leads to a jump of the temperature derivative of specific heat at the singular point. According to the terminology, introduced by Ehrenfest, a third-order transition occurs in the ideal Bose gas. In the paper of London (1938a) we read: "... Though actually the λ-point of helium resembles rather a second-order phase transition, it seems difficult not to imagine a connection with the condensation phenomena of the Bose–Einstein statistics ..."; and further: "... The different states of atoms may be described by eigenfunctions of a type similar to the electronic eigenfunctions which appear in Bloch's theory of metals. ... But in the present case we are obliged to apply Bose–Einstein statistics instead of Fermi statistics.... In his well-known papers Einstein has already discussed a peculiar condensation phenomenon of the 'Bose–Einstein' gas. According to this picture one fraction of atoms is condensed in the lowest energy state and the other occupies the excited, Bloch-like, states."

The paper of London (1938a) suggested to Tisza (1938) the idea of the so-called "two fluid theory" of helium II. "... A preliminary estimation shows that the atoms belonging to the lowest energy state do not take part in the dissipation of momentum. Thus the viscosity of the system is entirely due to the atoms in excited states." On the basis of this result Tisza (1938) tried to explain the previous experiments on helium II.

...(1) Measurements of the damping of an oscillating cylinder will show effect of atoms in excited states only....

(2) It is, however, not the viscosity coefficient μ which determines the flow velocity of helium II moving under influence of a pressure gradient though a capillary. In this case, the fraction of substance consisting of atoms in the lowest energy state will perform—like a superfluid liquid of viscosity $\mu \sim 0$—some sort of the capillary....

(3) The so-called fountain effect in helium II is the inverse process to (2). If one maintains a temperature difference between the ends of a capillary, a gradient in the density of excited atoms, n, and thus, in the

pressure is produced. In consequence, (a) the excited atoms will diffuse towards the colder end, and (b) the superfluid fraction of the liquid moves in the opposite direction. In the case of a wide tube, these currents must be equal and no resulting flow be observed. If, however, the capillary is sufficiently narrow, the rate of the process (a) becomes reduced and the temperature gradient causes a surplus convection current opposite to the heat flow....

Tisza, like initially Kapitza (1938), considered the motion of helium II as some kind of turbulent motion. The reason was that, as we stressed earlier, the velocity–pressure dependence is for helium II more similar to turbulent than to laminar motion.

During the following years Tisza (1940, 1947) developed his theory. The most important success was the prediction of the existence of temperature waves in helium II (1940). The theory, in the vicinity of absolute zero, gives for the velocity c_2 of waves the following temperature dependence $c_2 \sim \sqrt{\theta}$.

In 1941 Kapitza (1941a reprinted in Part 2 as paper E 10 on p. 114; 1941b) published his fundamental experimental papers allowing the possibility of a deeper understanding of the phenomenon of superfluidity giving a basis for the two-fluid theory, though when he wrote his papers, he was not familiar with this theory (see footnote to paper E10).

A two-fluid theory founded on a new theoretical basis, quite different from that of Tisza, was given by Landau (1941, reprinted in Part 2 as paper T1 on p. 191). Although the new theory is also called a two-fluid theory, its fundamental assumption is the possibility of the existence of two independent motions in helium II. These two motions occur without momentum transfer from one to another.

For the work on the theory of liquid helium L. D. Landau was awarded the Nobel Prize in Physics for 1962.

In his first paper Kapitza (1941a) established the hierarchy of the events in helium.

...Keesom and Keesom showed that liquid helium II in capillary tubes possesses an usually large heat conductivity which, by analogy to

> superconductivity in metals they named "superheat conductivity". In opposition to this view the author put forward a hypothesis which held that this abnormal heat conductivity was not due to some exceptional thermal property of helium II but to heat transferred by convection currents whose presence can be anticipated owing to the exceptionally high fluidity of helium II, and the author suggests that it should be named "superfluidity".... Therefore, we put forward the view that helium II must be regarded as a superfluid rather than as a "superheatconducting" liquid.

One of the problems investigated by Kapitza was the mechanism of the thermal conductivity in helium II, heat-induced motion. The experiments show that, if in the bulb (a Dewar vessel with heater, connected with the helium II reservoir by a capillary) heat is applied, there is a steady outflow of liquid from the vessel. As the vessel never empties it indicates the existence of an inflow into the bulb. The stream going out of the vessel transforms heat and has viscosity, in contrast to the opposite stream. In the language of the two-fluid model the first stream can be identified with the normal and the second one with the superfluid component. From this and further experiments one concluded that the superfluid component can interpenetrate through the superfluid one. This is at the present time considered to be the true mechanism of the thermal superconductivity discovered by Keesom and Keesom (1936).

Another fundamental experiment which helped to found the two-fluid model was done by Andronikashvili (1946, reprinted in Part 2 as paper E11 on p. 154). The density ratios of the normal and the superfluid components are functions of temperature. In the extreme cases $\theta = 0°K$ and $\theta = \theta_\lambda$ we have either only the superfluid density or only the normal component density different from zero respectively. The measurements of the ratios of these densities and therefore the demonstration of the existence of the two components was performed by Andronikashvili. A great number of thin discs, narrowly separated one from another suspended in a cryostat with helium II can perform torsional oscillations. The oscillating discs pull only the normal component along. Therefore the moment of inertia (discs + liquid trapped and moving with the discs) is a function of temperature, which was demonstrated experimentally.

The fact that helium exists as a liquid right down to absolute zero obviously cannot be explained in terms of the classical theory and indicates that the theory of helium II must be a miscroscopic one, based on quantum theory.

In the classical treatment, a liquid is described by the local quantities $\varrho(\boldsymbol{r},t)$, $\boldsymbol{j}(\boldsymbol{r}, t)$, $\theta(\boldsymbol{r}, t)$ or $(\varrho, \boldsymbol{v}, \theta)$. In the quantization of the motion of the liquid Landau considered suitable operators $\hat{\varrho}$, $\hat{\boldsymbol{j}}$ and $\hat{\boldsymbol{v}}$, the last defined by the formula

$$\hat{\boldsymbol{v}} = \frac{1}{2}\left(\frac{1}{\hat{\varrho}}\hat{\boldsymbol{j}}+\hat{\boldsymbol{j}}\frac{1}{\hat{\varrho}}\right). \tag{1}$$

He stated (without proof) the following commutation relations for the components of $\hat{\boldsymbol{v}}$

$$\hat{\boldsymbol{v}}(\boldsymbol{r}_1)_j\hat{\boldsymbol{v}}(\boldsymbol{r}_2)_k - \hat{\boldsymbol{v}}(\boldsymbol{r}_2)_k\hat{\boldsymbol{v}}(\boldsymbol{r}_1)_j = \frac{\hbar}{i}\,\delta({}_1\boldsymbol{r}-\boldsymbol{r}_2)\frac{1}{\hat{\varrho}(\boldsymbol{r}_1)}\,(\text{curl }\hat{\boldsymbol{v}})_{jk}. \tag{2}$$

The derivation of (2) was given by Ziman (1953). From the commutation relations it follows that curl $\hat{\boldsymbol{v}}$ does not commute with the Hamiltonian, i.e. is not conserved. There is one exception, namely, if curl $\hat{\boldsymbol{v}} = 0$ over the whole volume of the liquid. In this case, curl $\hat{\boldsymbol{v}}$ commutes with the operators $\hat{\varrho}$, $\hat{\boldsymbol{v}}$ and $\hat{H}$. The state for which it vanishes over the whole of the liquid is a stationary one and a state of potential motion.

By virtue of these considerations Landau stated that there is no continuous transition between the states of potential motion (curl $\boldsymbol{v} = 0$) and those of a vortex motion (curl $\boldsymbol{v} \neq 0$) of a quantum liquid. It means that between the lowest level of the potential flow energy spectrum and the lowest level of the vortex flow spectrum there must be a certain finite energy gap Δ.

In order to describe this situation, in his first version, Landau (1941) assumed that the energy spectrum describing liquid helium as a whole consists of two branches

$$\varepsilon(p) = cp, \qquad \varepsilon = \Delta + \frac{p^2}{2\mu}. \tag{3}$$

The first branch gave the energy-momentum dependence for sound quanta, phonons (c is the sound velocity). The second branch gave the energy of the "elementary excitations" which Landau regarded as characteristic for the vortex spectrum and following a suggestion of I. Tamm called "rotons". The roton spectrum is determined here by two empirical constants Δ and μ. The order of magnitude of Δ estimated by Landau is the same as of $k\theta_\lambda$ where θ_λ is the temperature of the λ-point.

Phonons like photons obey Bose statistics. Rotons obey probably also Bose statistics. Rotons can be treated as a gas only at low temperatures. Then Δ is large compared with $k_B\theta$, (k_B is BOLTZMANN'S constant) and there is practically no difference between the Bose and Fermi statistics

$$\left(\exp\frac{\Delta}{k_B\theta} \sim \exp\frac{k_B\theta_\lambda}{k_B\theta} \gg 1 \quad \text{because} \quad k_B\theta_\lambda \gg k_B\theta\right).$$

Therefore we can use for rotons the Boltzmann distribution.

The identification of the energy spectrum of helium II with the spectrum of phonons and rotons is the main idea behind the concept of elementary excitations. In this case it is very easy to obtain expressions for the free energy, the specific heat, and the entropy. From the above considerations it is clear that the contributions to these thermodynamic functions can be calculated independently for phonons and for rotons.

About the "elementary excitations", which are very often called "quasiparticles", Landau (1949) wrote:

> ...It follows unambiguously from quantum mechanics that for every slightly excited macroscopic system a concept can be introduced of "elementary excitations" which describe the collective motion of the

particles and which have certain energies ε and momenta $\boldsymbol{p}$. It is this assumption, indisputable in my opinion, which is the basis of the microscopic part of my theory. On the contrary every consideration of the motion of individual atoms in the system of strongly interacting particles is in contradiction with the first principles of quantum mechanics.

Landau stated that the existence of friction between liquid and walls is connected with the excitation of phonons or rotons in the liquid. For example, the motion of helium through a capillary "... must begin with the excitation of the inner movements in the layer in the liquid close to the walls of the tube, i.e. with the excitations of phonons in the liquid".

Landau proved, however, that for the assumed energy spectrum, for sufficiently small velocities (below the so-called critical velocity v_c) of helium the creation of phonons or rotons is energetically not possible and hence the liquid can perform a "superfluid" flow.

By considering the flow of helium through a capillary two coordinate systems are singled out: the system moving with the helium and the system at rest connected with the walls of the capillary (laboratory system). Let us assume that in helium flowing with a velocity $\boldsymbol{v}$ relative to the walls of a capillary an excitation is created with energy ε and momentum $\boldsymbol{p}$. The creation of an excitation changes, of course, the energy of the liquid in the laboratory system. The creation is possible, if the change of energy leads to a lowering of the total energy. In order to examine the situation more accurately and express the final result in terms of $\boldsymbol{v}$ we need formulae relating the momenta and energies in the system moving with the helium and in the laboratory system. Consider a mass M of helium flowing with velocity $\boldsymbol{v}$ in the laboratory system. In the moving system the liquid has zero momentum and an energy E_0. In the laboratory system the momentum is $M\boldsymbol{v}$ and the energy $E_0+\frac{1}{2}Mv^2$. If in the liquid an excitation is created with energy ε and momentum $\boldsymbol{p}$, then in the system moving with liquid the momentum is $\boldsymbol{p}$ and energy $E_0+\varepsilon$. In the laboratory system the momentum is $\boldsymbol{p}+M\boldsymbol{v}$ and the energy $E_0+\varepsilon+\frac{1}{2}Mv^2+(\boldsymbol{p}\cdot\boldsymbol{v})$. Hence the change of energy in laboratory system due to the creation of an excitation is equal to $\varepsilon+(\boldsymbol{p}\cdot\boldsymbol{v})$. It is expressed in terms of the excitation energy $\varepsilon(p)$ and the velocity $\boldsymbol{v}$. The change can be negative, lowering the total energy of liquid, if

$$\varepsilon(p)+(\boldsymbol{p}\cdot\boldsymbol{v}) < 0. \tag{4}$$

To fulfil this condition the velocity v must be greater than the critical velocity

$$v_c \equiv \left(\frac{\varepsilon(p)}{p}\right)_{\min}. \tag{5}$$

It means that for velocities $v < v_c$ no excitations can be created and the liquid remains superfluid.

For the excitation of a phonon v_c, calculated from (5), is about 25,000 cm/sec and for the excitation of a roton (with $\varepsilon(p)$ given by (8)) v_c is about 8000 cm/sec. The comparison with the experimental data $v_c \sim 70$ cm/sec, shows that there is a great difference between the theoretical and experimental results. This discrepancy needed an explanation which was given in the further development of the theory.

The idea of a critical velocity was first suggested by Daunt and Mendelssohn (1939b) in order to explain observations of the helium film flow out of beakers. In such films the superfluid moves due to the gravitational potential difference. However, the observed velocity was finite and independent of the potential difference. For the superfluid moving "without friction" such a velocity seems to be infinite. Daunt and Mendelssohn advanced the hypothesis that the velocity cannot increase arbitrarily because of the existence of a certain "critical velocity" beyond which helium is not superfluid. From the measurements of the thickness of such films one estimated the upper limit of v_c to be about 70 cm/sec.

Now we shall briefly discuss the considerations of Landau (1941) which led him to the two-fluid model. Consider helium in a vessel rotating at a small angular velocity $\boldsymbol{\Omega}$. The total angular momentum is given by the formula

$$M = \frac{1}{(2\pi)^3} \int\int \frac{m\, d^3p\, d^3r}{\exp\left[\dfrac{\varepsilon - (\boldsymbol{\Omega} \cdot \boldsymbol{m})}{k_B \theta}\right] - 1}, \tag{6}$$

where $\boldsymbol{m}$ is the angular momentum of a phonon or roton and ε the appropriate energy. In formula (6) the ratio of $\boldsymbol{M}$ to $\boldsymbol{\Omega}$ gives the momentum of inertia of the liquid. Therefore at $\theta = 0$, in the absence of elementary excitations, $\boldsymbol{M}$ vanishes and automatically the moment of inertia is equal to zero. At higher temperatures, if some excitations are present, $\boldsymbol{M}$ is different from zero. However, the moment of inertia is much smaller than that corresponding to the total mass of helium. From this picture it follows that only a part of the liquid connected with the elementary excitations should take part in the rotation.

Through these considerations Landau arrived at the two-fluid model. It is quite different from Tisza's model.

The intuition, however, which brought Landau to the two-fluid model was, as it later appeared, based on a "Gedanken experiment". Real experiments by Osborne (1950) showed that in a uniformly rotating vessel both the normal and superfluid components are in rotation.

This is the second discrepancy of Landau's theory with experiment in addition to the critical velocity predicted by this theory which is too high.

It turned out that these two difficulties have a common origin in the assumption that curl $\boldsymbol{v}_s \equiv 0$ in the whole volume of the liquid. We shall explain this problem more exactly later on.

Landau (1941) noted that

> ...It must be stressed that when we look upon helium as a mixture of two liquids this is no more than a method of expression, convenient for describing phenomena which take place in helium II. Like every description of quantum phenomena in classical terms it is not quite adequate. Actually one must say that in a quantum liquid two movements can exist simulta-

neously, each of which is connected with its own "effective mass" (so that the sum of both these masses equals the total real mass of the liquid). One of these movements is normal, i.e. possesses the same properties as the movements of usual liquids; the other is superfluid. Both these motions take place without a transfer of the momentum from one to the other*. We particularly emphasize that there is no division of the real particles of the liquid into "superfluid" and "normal" ones here.† In a certain sense one can speak of "superfluid" and "normal" masses of liquid as of masses connected with the two simultaneously possible movements, but this by no means signifies the possibility of a real division of the liquid into two parts.

From the considerations of Landau we see that the rotating part of liquid connected with the elementary excitations behaves as an ordinary liquid, this part has viscosity. The stationary part does not show viscosity and is superfluid. Therefore it is necessary to introduce some new local quantities characteristic for the superfluid. Specifically we introduce the density ϱ_n and velocity $\boldsymbol{v}_n$ of the normal component and the density ϱ_s and velocity $\boldsymbol{v}_s$ of the superfluid component. Now it is possible to derive the hydrodynamic equations describing two independent motions in helium II. This was done by Landau (1941). "... It turns out that the hydrodynamic equations with the two velocities $\boldsymbol{v}_s$ and $\boldsymbol{v}_n$ can be obtained absolutely unambiguously starting from one condition only that they satisfy all the conservation laws."

These conservation laws are for mass, momentum and entropy. Additionally we consider a new equation for the acceleration $d\boldsymbol{v}_s/dt$. The superfluid velocity is assumed to be a potential velocity.

The calculation of the thermodynamic functions performed by Landau was based on the quantum mechanical energy spectrum of the system. Since the form of the spectrum was postulated and not derived from first principles this level of the theory can be

* The further development of the theory shows that the creation of the quantized vortex lines in the superfluid component and the scattering of phonons and rotons by these lines lead to "mutual friction".

† This remark is especially important in order to distinguish between Tisza's and Landau's approach to the two-fluid theory.

called quasi-microscopic. The derivation of the fundamental equations of hydrodynamics of helium II was given by Landau on the macroscopic level. Recently these equations were obtained by Bogoliubov (1963) and independently by Hohenberg and Martin (1965) from the microscopic theory. Later we shall mention Bogoliubov's method.

After linearization of Landau's equations it is possible to obtain two wave equations describing the propagation of two independent wave motions in helium II. They were called by Landau first and second sound. The velocity of first sound c_1 turned out to be the well-known velocity of ordinary sound. But the velocity of second sound c_2 was something specific for helium II. Landau obtained the temperature dependence for c_1, c_2 especially for $\theta \to 0$.

We mentioned earlier that the prediction of the temperature waves was a great success of Tisza's theory. Tisza at once understood the nature of the second wave process in helium. He found at the same time a formula describing the temperature dependence of c_2. From this formula it follows that at absolute zero c_2 vanishes. This prediction is quite different from Landau's that at $\theta = 0$: $c_2 = (1/3)c_1\ 1/\sqrt{3}c_1 \neq 0$. The decision of which theory is in better agreement with experimental data was very difficult because measurements of c_2 were performed only relatively recently (Atkins and Osborne, 1950; Peshkov, 1960). These experiments showed the superiority of Landau's theory, though Peshkov emphasizes that "in experimental conditions accessible at present it seems doubtful to speak about second sound at temperatures below 0·5°K."

In Tisza's theory (1947) the elementary excitations "correspond to helium atoms in translational Bloch type states" and Debye phonons "are associated with the liquid as a whole".

The wave function ψ of helium has the following form:

$$\psi = \psi_0(1, \ldots, N)\psi_{\text{phon}}\psi_{\text{gas}},$$

$$\psi_{\text{phon}} = \prod \psi(u_i), \quad \psi_{\text{gas}} = \sum_{\text{permut}} \prod_{i=1}^{N_{ex}} \exp\,[i(\boldsymbol{k}_i\boldsymbol{r}_i)] \tag{7}$$

where N is the total number of molecules, N_{ex} the number of excited molecules, u_i the longitudinal normal modes involving the coordinates of all the atoms. The function ψ_0 $(1, \ldots, N)$ is the eigenfunction of the lowest state. It depends on the coordinates of all the atoms. The function ψ_{gas} depends only on the coordinates of the excited atoms.

"... In Landau's theory the excited atoms are replaced by 'rotons' ...but Landau defined the normal component as a 'gas of phonons and rotons', leaving the superfluid component with rigorously vanishing entropy." (Tisza 1947).

Landau (1949) has discussed and criticized the approach of Tisza in the article "On the theory of superfluidity":

> "...On the contrary, every consideration of the motion of individual atoms in the system of strongly interacting particles is in contradiction with the first principles of quantum mechanics."
>
> "The hydrodynamic equations given by Tisza are, in my opinion, quite unsatisfactory. It is easy to see that in their exact form they even violate the conservation laws!"
>
> "Tisza concludes that the viscosity must decrease with decreasing temperature ...recent viscosity measurements by E. Andronikashvili, which correspondingly are in conflict with Tisza's considerations."

After the prediction of second sound by Landau in 1941 the first attempts of Shalnikov and Sokolov to generate second sound and measure its velocity were unsuccessful. The reason was that they tried to excite second sound purely by mechanical methods (by oscillating piezo-quartz).

In 1944 Lifshitz (1944, reprinted in Part 2 as paper T2 on p. 234) gave the theoretical examination of the effectiveness of various methods of exciting second sound. By "mechanical" excitations the ratio of intensities for second and first sound is equal to $I_2/I_1 \sim \alpha^2\theta(c_2^3/c_1)$, where α is the thermal expansion coefficient, e.g. at $\theta \sim 2^2$K the ratio I_2/I_1 is very small, of the order of 2×10^{-6}. If the second sound source consists in periodical variations of the temperature we have

$$I_2/I_1 \sim \frac{c}{\theta\alpha^2 c_1 c_2}$$

where $c \approx c_v \approx c_p$ is the specific heat. For $\theta \sim 2°K$ we have now I_2/I_1 of the order 5×10^3. In 1944 Peshkov (1944, reprinted in Part 2 as paper E12 on p. 166) produced and observed standing thermal waves in helium II. Periodic temperature fluctuations were created by a plane electric heater (electric current frequency 100–10,000 Hz). For measurements below 2·2°K Peshkov found that at 1·65°K c_2 reached a maximum of 20·3 m/sec and then decreased to a minimum of 18·4 m/sec at 1·1°K. Further measurements by Atkins and Osborne (1950) showed that c_2 rose again to a value of about 150 m/sec on a further lowering of the temperature. The first sound velocity increased below 2·2°K from 220 m/sec to 250 m/sec. As we mentioned before, these results are in agreement with the predictions of Landau's theory as $\theta \to 0$. However, for finite temperatures there was a discrepancy between theory and experiment, e.g. for $\theta = 1{\cdot}6°K$ Peshkov obtained $c_2 = 19$ m/sec whereas Landau's theory predicted $c_2 = 25$ m/sec. An analysis of the data of this experiment and the result of a paper by Bogoliubov (1947, which is reprinted as paper T4 on p. 247) led Landau (1947, reprinted in Part 2 as paper T3 on p. 243) to a new assumption about the energy

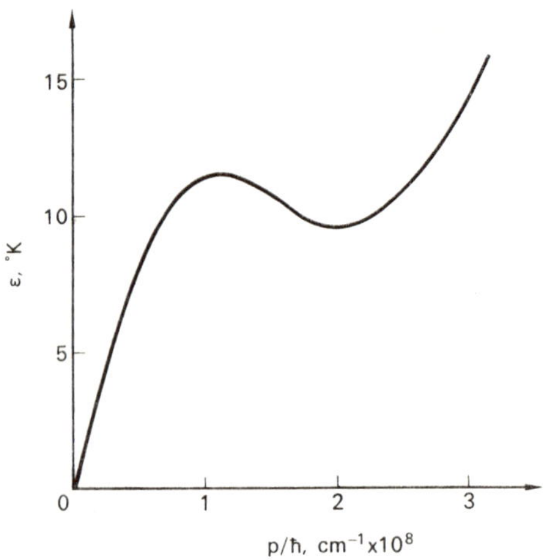

FIG. 2. The energy spectrum of helium II (after Landau, 1947).

spectrum of helium II. It is described by one dispersion curve $\varepsilon = \varepsilon(p)$ (see Fig. 2; compare equation (11) below). For small momenta it is linear in p like the phonon branch in the first version. Then when $p/\hbar = p_0/\hbar \sim 1/a$ (where a is the mean atomic distance) $\varepsilon(p)$ has a minimum $\varepsilon(p_0) = \Delta$. This minimum is equal to the energy gap which in the first version divided the potential motion energy from the vortex motion energy. In the vicinity of p_0 we have

$$\varepsilon(p) = \Delta + \frac{(p-p_0)^2}{2\mu}. \tag{8}$$

"With such a spectrum it is of course impossible to speak strictly of rotons and phonons as of qualitatively different types of elementary excitations. It would be more correct to speak simply of the long wave (small p) and short wave (p in the neighbourhood of p_0) excitations" (Landau, 1947). One can speak also about the phonon or roton part of the dispersion curve.

The contribution to the thermodynamic functions can be calculated separately for the phonon and roton part of the curve. The phonon contribution to the free energy, entropy, specific heat (per unit mass) and the density of the normal component are, respectively,

$$\begin{aligned} F_{\mathrm{ph}} &= -\frac{\pi^2 k_{\mathrm{B}}^4}{90\hbar^3 c^3}\,\theta^4, & S_{\mathrm{ph}} &= \frac{2\pi^2 k_{\mathrm{B}}^4}{45\varrho\hbar^3 c^3}\,\theta^3, \\ c_{\mathrm{ph}} &= \frac{2\pi^2 k_{\mathrm{B}}^4}{15\varrho\hbar^3 c^3}\,\theta^3, & (\varrho_n)_{\mathrm{ph}} &= \frac{2\pi^2 k_{\mathrm{B}}^4}{45\hbar^3 c^5}\,\theta^4. \end{aligned} \tag{9}$$

The thermodynamic functions for the "roton gas" are

$$\begin{aligned} F_r &= -\frac{2\mu^{1/2}(k_{\mathrm{B}}\theta)^{3/2}p_0^2}{(2\pi)^{3/2}\varrho\hbar^3}\, e^{-\Delta/k_{\mathrm{B}}\theta} \\ S_r &= \frac{2(k_{\mathrm{B}}\mu)^{1/2}p_0^2\Delta}{(2\pi)^{3/2}\varrho\theta^{1/2}\hbar^3}\left(1+\frac{3k_{\mathrm{B}}\theta}{2\Delta}\right) e^{-\Delta/k_{\mathrm{B}}\theta-\Delta/k_{\mathrm{B}}\theta}, \\ c_r &= \frac{2\mu^{1/2}p_0^2\Delta^2}{(2\pi)^3\varrho k_{\mathrm{B}}^{1/2}\theta^{3/2}\hbar^3}\left[1+\frac{k_{\mathrm{B}}\theta}{\Delta}+\frac{3}{4}\left(\frac{k_B\theta}{\Delta}\right)^2\right] e^{-\Delta/k_{\mathrm{B}}\theta}, \\ (\varrho_n)_r &= \frac{2\mu^{1/2}p_0^4}{3(2\pi)^{3/2}(k_{\mathrm{B}}\theta)^{1/2}\hbar^3}\, e^{-\Delta/k_{\mathrm{B}}\theta}. \end{aligned} \tag{10}$$

The density of the normal component can be defined with the help of a formula similar to (6). We obtain an expression for the total momentum $\boldsymbol{P}$ per unit volume by replacing in (6) $\boldsymbol{m}$ (the angular momentum of a phonon or roton) by $\boldsymbol{p}$ (the linear momentum of a phonon or roton), $(\boldsymbol{\Omega}\cdot\boldsymbol{m})$ by $(\boldsymbol{v}\cdot\boldsymbol{p})$ where $\boldsymbol{v}=\boldsymbol{v}_n$, $\boldsymbol{v}_s=0$ and omitting the integration over the volume. For small $\boldsymbol{v}$ this integral expression can be expanded in powers of $(\boldsymbol{v}\cdot\boldsymbol{p})$. The ratio of $\boldsymbol{P}$ to $\boldsymbol{v}$ is just ϱ_n. If we take for ε the phonon energy given by (3) we obtain the phonon density $(\varrho_n)_{\text{ph}}$. If we take for ε the roton energy given by (8) and use the Boltzmann distribution we obtain the roton density $(\varrho_n)_r$. In both cases we can extend the integration limits to infinity because of the fast convergence of the integrands.

In the roton spectrum there appear three phenomenological constants Δ, p_0, and μ. They enter also into the roton part of the thermodynamic functions. From the thermodynamic data (on second sound) Landau (1947) estimated that

$$\frac{\Delta}{k_B} = 9{\cdot}6°\text{K}, \quad \frac{p_0}{\hbar} = 1{\cdot}95 \quad (\text{Å}^{-1}),$$

$$\frac{\mu}{m_{\text{He}}} = 0{\cdot}77.$$

The parameters of the Landau spectrum can be determined also by neutron scattering experiments. Mono-energetic neutrons scatter inelastically by rotons or phonons. By measuring the initial and final momenta of neutrons as well as the angle of scattering it is possible to reproduce the whole spectrum $\varepsilon(p)$ of elementary excitations in helium II (Yarnell *et al.*, 1959; Henshaw and Woods, 1961). From neutron scattering experiments we have

$$\frac{\Delta}{k_B} = 8{\cdot}6°\text{K}, \quad \frac{p_0}{\hbar} = 1{\cdot}9 \quad (\text{Å}^{-1}),$$

$$\frac{\mu}{m_{\text{He}}} = 0{\cdot}16$$

Discussing Landau's theory of superfluidity London (1954) wrote: "Landau rejected the approach from the side of gas statistics, arguing that a liquid is generally much more like a solid than a gas,* and he attempted to put his two-fluid theory on an essentially different basis, one in which the Bose–Einstein condensation plays no role at all."

The connection between the microscopic theory of superfluidity and the Bose-Einstein condensation was demonstrated by Bogoliubov (1947, reprinted in Part 2 as paper T4 on p. 247).

> The object of this paper is an attempt to construct a consistent molecular theory explaining the phenomenon of superfluidity without assumptions concerning the structure of the energy spectrum.... In our theory the existence and properties of the elementary excitations follow directly from the basic equations describing the Bose–Einstein condensation of non-perfect gases.... The particles of a degenerate perfect Bose–Einstein gas in the ground state cannot possess the property of superfluidity, since nothing prevents them from exchanging their momenta with excited particles colliding with them, and therefore, from friction in their movement through the liquid.†
>
> In the present paper we try to overcome this difficulty and show that under certain conditions the "degenerate condensate" of a "nearly perfect" Bose–Einstein gas can move without any friction with respect to the elementary excitations, with an arbitrary, sufficiently small velocity.

Bogoliubov considered a nearly perfect Bose–Einstein gas with "degenerate condensate." Condensation occurs into a ground state. The number of molecules into the condensate N_0 is very large as compared with unity. If it is assumed that in the system there is a state with one macroscopically large occupation number, then a decrease or increase by one particle can be neglected. Hence the creation and annihilation operators a_0^+, a_0, corresponding to this state, can be treated as c-numbers. It makes it possible to diagonalize the Hamiltonian of the "nearly perfect" Bose–Einstein

* The view that a liquid should be considered essentially a solid which is deprived of its long-range order has been developed systematically in particular by J. Frenkel (1946).

† This fragment contains a critique of Tisza's approach.

gas. The diagonalized Hamiltonian describes "a perfect Bose–Einstein gas of certain 'quasi-particles' representing the elementary excitations, which cannot be identified with the individual molecules" (Bogoliubov, 1947). The eigenvalues of the Hamiltonian give the dispersion curve describing the energy-momentum dependence for quasiparticles

$$\varepsilon(k) = \sqrt{\left(\frac{k^2\nu(k)}{mv} + \frac{k^4}{4m^2}\right)}, \tag{11}$$

where $\nu(k)$ is the Fourier component of the interaction potential energy of the particles and v the volume per particle. For small momenta the dependence is linear, $\varepsilon(p) = cp$, typical for phonons. For large momenta the dependence is quadratic, $\varepsilon(p) = p^2/2m$, like the energy of a free particle. According to the important remark by Bogoliubov that "$\nu(k)$ tends to zero with increasing k" the derivative $d\nu(k)/dk$ is then negative, and it is possible to have $d\varepsilon(k)/dk = 0$. For intermediate momenta, the simplest shape of the energy–momentum curve is that given by Fig. 2.

The other important result of the paper of Bogoliubov is the proof that at absolute zero the condensate is only part of the superfluid: "We thus see that even at $\theta = 0$ only a fraction of molecules possesses momenta which are exactly zero, the rest being continuously distributed over the whole momentum spectrum."

Although the interaction between helium molecules is comparatively weak, liquid helium cannot be considered as a "nearly perfect" Bose–Einstein gas. So the problem of the justification of the Landau spectrum on the basis of a microscopic theory remained open. This important problem was undertaken by Feynman (1953, 1954).

> Landau has in fact, tried to obtain some justification for the spectrum from a study of quantum hydrodynamics. This is not a completely atomic approach. One attempts to describe the liquid by a few quantities such as density and current or velocity. Then one makes these quantities operators with resonable commutation relations, and tries to find excitation energies of the fluid. The problem has not been analysed in sufficient detail to

establish the energy spectrum.... However, it is possible from first principles to see why there are no other excitations but those supposed by Landau and why the energy spectrum of these excitations has, qualitatively, the form which he supposed. (Feynman, 1955, reprinted in Part 2 as paper T5 on p. 268).

Feynman considered the most important properties of the wave function ψ describing an ensemble of N Bose particles. The amplitude ψ is a function of $3N$ variables $r_j^{(\alpha)}$, $\alpha = 1, 2, 3$. Some configurations of atoms are more, some less probable. The realization of a configuration depends on the interaction energy of the helium atoms. For example, at distances smaller than 2·7 Å the interaction is strongly repulsive. Hence the configurations in which atoms are closer than 2·7 Å have a very low probability. "To summarize, the ground state function is large for any configuration in which the atoms are well spaced from each other at near constant average density."

Classically sound vibrations can be described in terms of normal coordinates which obey the harmonic oscillator equation. The appropriate normal coordinates for the sound wave description are the Fourier components ϱ_k of the number density $\varrho(\boldsymbol{r})$. For the quantum mechanical density operator

$$\varrho(\boldsymbol{r}) = \sum_j \delta(\boldsymbol{r}-\boldsymbol{r}_j)$$

the Fourier components have the form

$$\hat{\varrho}_k = \sum_j [\exp i(\boldsymbol{k}\cdot\boldsymbol{r}_j)]. \tag{12}$$

The excited state wave function ψ_{ph} describing density fluctuations was assumed by Feynman by analogy to the first excited state wave function of the quantum mechanical oscillator to be

$$\psi_{\text{ph}} = \hat{\varrho}_k \Phi = \sum_j \exp [i(\boldsymbol{k}\cdot\boldsymbol{r}_j)]\Phi. \tag{13}$$

where Φ is the ground state function.

"The central problem is to see why no states other than phonons can have such low energies. We try to construct the wave function ψ of an excitation which should be as low in energy as possible and yet not represent a phonon."

The function ψ must be orthogonal to the ground state function Φ and to all the phonon states. Since Φ must be positive (has no nodes) ψ must be positive for half of the configurations and negative for the other half in order to fulfil the orthogonality condition. The only important configurations are those of roughly uniform density.

We are interested in the energies of the system which are as low as possible, therefore ψ must vary slowly in space. In the opposite case, the gradient of ψ should be too large, giving a large kinetic energy. Hence the two configurations for which ψ has opposite signs must be obtained one from each other by displacements of the atoms over large distances. In such a case grad ψ should be small. For the configuration of a roughly uniform density the distance between atoms in "positive" (or "negative") configurations are of about of atomic separation. These (+) and (—) configurations are especially different if the sites of atoms in "negative" configurations can fall between the atoms in "positive" configurations. In this case the distances between the nearest neighbours of two configurations are equal to half the atomic separation.

Consider a displacement of an atom 1 from a "positive" configuration, when it is at $\boldsymbol{r}_1^{(+)}$, to its position $\boldsymbol{r}_1^{(-)}$ in a "negative" configuration. This new position is among the atoms of the "positive" configuration, hence the neighbours of 1 are lying no further away than half the atomic separation. A displacement of the atom is connected with a density variation. In order to compensate this variation we must displace the nearest neighbour of $\boldsymbol{r}_1^{(-)}$, situated at $\boldsymbol{r}_2^{(+)}$ and belonging to the "positive" configuration, to the point $\boldsymbol{r}_2^{(-)}$ which is a next neighbour of $\boldsymbol{r}_1^{(+)}$

$$(\boldsymbol{r}_1^{(+)} \to \boldsymbol{r}_1^{(-)}, \quad \boldsymbol{r}_2^{(+)} \to \boldsymbol{r}_2^{(-)}; \quad \boldsymbol{r}_1^{(-)} \sim \boldsymbol{r}_2^{(-)}, \quad \boldsymbol{r}_1^{(-)} \sim \boldsymbol{r}_2^{(+)}).$$

The position $r_2^{(-)}$ lies between the atoms forming a "positive" configuration. Therefore the new position $r_2^{(-)}$ of the atom 2 is no further away from the initial position $r_1^{(+)}$ of the atom 1 than half an atomic separation. This simplified example illustrates that the displacements of atoms can take place practically without density variations. The described displacement, which we assume to be large, is equivalent to the changing of a "positive" configuration to a "negative" one. Hence ψ changes sign i.e. its change is large. In our considerations, however, we did not make use of the principle of the identity of the atoms in a Bose ensemble. The function must be symmetrized, i.e. remain unchanged under simple permutations of atoms. Then, ψ cannot change if atom 1, situated at $r_1^{(+)}$, and atom 2, situated at $r_2^{(+)}$, interchange their positions ($b_1^{(+)} \rightleftarrows b_2^{(+)}$). Hence the change of a (+) to a (−) configuration can be described as follows:

$$r_1^{(+)} \rightleftarrows r_2^{(+)},\ r_2^{(+)} \to r_1^{(-)},\ r_1^{(+)} \to r_2^{(-)};\ r_2^{(+)} \sim r_1^{(-)},\ r_1^{(+)} \sim r_2^{(-)}.$$

Therefore, for the (+) and (−) configurations which in fact are spatially close to one another, the function ψ undergoes a rapid change. These configurations describe states with practically the same density. We see, therefore, that because of the Bose statistics properties the gradient of the function ψ which describe non-phonon states, cannot be small.

Feynman identified the wave function related to the special configurations described above, with the wave function of the excited roton state. He put ψ_{rot} in the form

$$\psi_{rot} = \sum_j f(r_j)\Phi = F\Phi. \tag{14}$$

The presence of Φ in (14) guarantees that ψ_{rot} does not describe configurations in which atoms overlap. The form of the function $f(r)$ was found from a variational procedure by minimizing the expression

$$\varepsilon = \frac{\int \psi_{rot}^* \hat{H} \psi_{rot}\, d\tau}{\int \psi_{rot}^* \psi_{rot}\, d\tau}. \tag{15}$$

It gave for F

$$F = \sum_j \exp\,[i(\boldsymbol{k}\cdot\boldsymbol{r}_j)]. \tag{16}$$

For the energy Feynman obtained the very important formula

$$\varepsilon(\boldsymbol{p}) = \frac{p^2}{2mS(\boldsymbol{p})}, \tag{17}$$

$$S(\boldsymbol{p}) = \int \rho(\boldsymbol{r}) \exp\left[i\left(\frac{\boldsymbol{p}}{\hbar}\cdot\boldsymbol{r}\right)\right] d^3\boldsymbol{r}$$

The function $\rho(\boldsymbol{r})$ is the two-atom correlation function giving the probability (per unit volume, in the ground state) to find an atom at a distance $\boldsymbol{r}$ from a given atom. The Fourier transform $S(\boldsymbol{p})$ is called the liquid "structure factor". It can be found from X-ray scattering experiments (Reekie and Hutchinson, 1953; Beaumont and Reekie, 1955; Gordon, Shaw, and Daunt, 1958). The measurements of $S(\boldsymbol{p})$ can be performed only at finite temperatures whereas in formula (17) the "structure factor" appears for $\theta = 0$. Feynman and Cohen (1956) analysed the form of $S(p)$ theoretically. This was possible only using a special assumption justified for small p. This case corresponds to the wavelength region in which the liquid can be treated as a continuous medium. For $p \sim 0$ they obtained

$$S(p) = \frac{p}{2mc} \tag{18}$$

where c is the sound velocity. From (17) and (18) we have

$$\varepsilon(p) = cp, \tag{19}$$

i.e. the phonon spectrum, which was fully confirmed by experiments. Further theoretical considerations by Feynman and Cohen showed that for $\theta \neq 0$ and small p the function $S(p)$ does not only depend on p but is also a function of temperature. The experiments by Gordon, Shaw and Daunt (1958) confirmed this result. So for a small p, $S(p)$ can be obtained, for $\theta = 0$, only from theoretical considerations. For a larger p, especially in the vicinity $p/\hbar \sim 2\,\text{Å}^{-1}$, the model of a continuous medium is not valid and $S(p)$ must be taken from the experimental data. In order to reproduce the form of $S(p)$ the experimental data are taken from the paper by Reekie and Hutchinson (1953) (the more recent data are given in a paper by Beaumont and Reekie (1955)). All experimental measurements give a sharp maximum of the $S = S(k)$ curve for $k \sim 2\,\text{Å}^{-1}$. It leads to a minimum in the dispersion curve $\varepsilon(p)$, characteristic for the roton part of the spectrum. Although the qualitative agreement is satisfactory quantitatively it is bad. The experimental value of $\varepsilon(p)$ at the point $p/\hbar \sim 1{\cdot}8\,\text{Å}^{-1}$ is of about 8·7°K whereas formula (17) gives a value of 19°K. One reason for this discrepancy could be that in (17) the experimental values for $S(p)$ are used referring to finite temperatures. Experimental data, however, especially from the paper by Gordon, Shaw and Daunt (1958), show that for k in the vicinity of $2\,\text{Å}^{-1}$ the function $S(k)$ is practically independent of temperature. The second reason for the discrepancy is that the form of the function F in (14) is not satisfactory. In fact, Feynman (1954) found that the wave function (14), (16) does not satisfy the equation of continuity.

> We shall analyse this for the case of a single excitation of nearly definite momentum in the form of a large wave packet.... It is true that the distance of influence in ϱ may not be very small because of the correlations in the sound field. That is, the excitation produces a small strain in the fluid which makes a field of stress in the vicinity. Such fields provide a mechanism of interaction between excitations.... In a more detailed analysis such effects should be taken into account. Here we proceed to a first approximation and neglect them. To the approximation of neglecting compressibility, then, we find $\varrho(\boldsymbol{a}) = \varrho_0$; the presence of an excitation does not change the fluid density.
>
> Thus we picture an excitation in the form of a drifting wave packet as carrying a total current $\hbar\boldsymbol{k}/m$, and drifting (if $\boldsymbol{v} = 0$) at the group velocity $\boldsymbol{v}_g = \partial\varepsilon/\partial\boldsymbol{k}$, but as not appreciably alternating the density.
>
> This clearly violates the conservation of matter.
>
> The problem can be resolved by considering more complicated functions representing interaction of the excitation with the flow of fluid in its surroundings. One way the current could be conserved would be to have a general return flow of fluid in the region outside the packet. We therefore

try the solution

$$\psi = \sum_i g(r_i) \exp [i(\boldsymbol{k}\cdot\boldsymbol{r}_i)] \exp [i \sum_j s(\boldsymbol{r}_j)]\Phi \tag{20}$$

with the hope of finding an s which produces a velocity distribution $\boldsymbol{v} = \hbar\nabla s/m$ which shows such a reverse flow (Feynman, 1954).

In this case the total current is of the form $\boldsymbol{J} = \boldsymbol{j} + \varrho_0 (\hbar/m)\nabla s$ and now the current $\boldsymbol{J}$ is conserved:

$$(\nabla\cdot[\boldsymbol{j} + \varrho_0(\hbar/m)\nabla s]) = 0. \tag{21}$$

The current $\varrho(\hbar/m)\nabla s$ describes just the backflow.

The velocity potential was found by Feynman and Cohen (1956). It is represented by a function of the form $g(\boldsymbol{r}_i - \boldsymbol{r}_j)$ different from $s(\boldsymbol{r}_j)$ proposed by Feynman.

Feynman and Cohen took the wave function ψ in the form

$$\psi = \sum_\alpha \exp [i(\boldsymbol{k}\cdot\boldsymbol{r}_\alpha)] \exp [i \sum_\beta g(\boldsymbol{r}_\beta - \boldsymbol{r}_\alpha)] \tag{22}$$

with

$$g(\boldsymbol{r}) = A\frac{(\boldsymbol{k}\cdot\boldsymbol{r})}{r^3}.$$

This form of $g(\boldsymbol{r})$ was suggested to the authors after consideration of the motion of a foreign atom through the fluid. If the atom is treated classically as a sphere, the velocity potential at great distances from the sphere, describing the backflow around it, is just given by (22) with $A = -3{\cdot}6$ Å^3.

The function $g(\boldsymbol{r})$ has the form of a velocity potential for a dipole flow. The foreign atom (impurity, e. g. ^{3}He atom) was assumed to have the same mass as a helium atom. The wave function describing the pattern of backflow in the case of an impurity is of the form (22) (only without a summation over α). The function (22) is an antisymmetrized eigenfunction of momentum and leads to the conservation of local current. From the variational principle Feynman and Cohen found $A = -3{\cdot}8$ Å^3 which is close to the classical value and suggests that the choice of (22) was adequate. In terms of (22) and the experimental data of Reekie and Hutchinson (1953), the authors found that Δ/k_B is about 11·5°K which is in better agreement with experiment than results following from (17).

The fluid velocity in the coordinate system connected with a particular atom in the helium has the form

$$v(\boldsymbol{r}) = -\left(\frac{1}{m}\right) \wp(\boldsymbol{r}) \nabla g(\boldsymbol{r}) + \dots \tag{23}$$

where $\wp(\boldsymbol{r})$ is the pair correlation function between two helium atoms (see Chester, 1963). The further terms in (23) describe the more complicated correlation effects.

The first term in (23) is the same as for an impurity atom. For $\wp(\boldsymbol{r}) = 1$, which is realized at large r, the first term is the same as the classical velocity field, similar to a velocity field of a small smoke ring. "The roton state represented by the function (21)* can be described roughly classically as a vortex ring of such a small radius that only one atom can pass through the center.... The ring does not drift forward as a large smoke ring, because as it is as small as possible there is no force tending to shrink it", (Feynman and Cohen, 1956).

"Some years ago, the Yale theoretician L. Onsager, when asked to define the roton, called it 'the ghost of a vanishing vortex ring'," (Lane, 1962).

Another interpretation is given by de Boer (1963): "The rotons then lost completely their 'vortex' or rotation character and should instead be considered as short-wavelength longitudinal elastic modes, which are completely degenerated because the wavelength is of the order of magnitude of the average distance between neighbouring atoms." The comparison of the dispersion curve for longitudinal elastic waves in a crystal lattice with the energy momentum curve for excitations in liquid helium was the basis

* In our text (22).

for the above statements. "Now the interesting point is that this dispersion curve is rather similar to what one could expect to occur in any condensed substance."

In three-dimensional crystals it is difficult to obtain the dispersion curve for longitudinal elastic waves because they can propagate only in some special directions and the dispersion curves depend on direction. In the article by de Boer (1963) one dispersion curve given after averaging over the curve has a maximum for $k_{max} \sim 4{\cdot}5a$, where a is the mean distance between molecules. The k_{max} is close to the Debye momentum $k_D = 3{\cdot}9a$. For k exceeding k_{max} the curve goes to a flat minimum. The energy–momentum curve for liquid helium has a shape very similar to the one described. It has a maximum at k_D and then a minimum at $k \sim 2$ Å^{-1}. On increasing k the wavelength of the elastic waves decreases. For $k \sim 1/a \sim 2$Å^{-1} the wavelengths become extremely short, of the order of a. Below these wavelengths it is impossible to speak about collective oscillations. Hence one can speak only about completely degenerate short-wavelength longitudinal elastic modes.

In an article by Chester (1963) a roton is associated with a ^{4}He atom with modified mass.

> The conjecture I want to put forward is that in the region of the minimum of the energy spectrum we are dealing with the motion of a single helium atom moving fast through the others.... The picture we have formed so far is as follows. From $|\boldsymbol{k}| = 0$ up to $|\boldsymbol{k}| \simeq 1$ Å^{-1} the motions in the fluid consist of longitudinal density waves.... On the other hand for $k \simeq 2$ Å^{-1} we find that there is essentially no local density fluctuation and we have a single particle propagating fast through the liquid. The increase in effective mass of this atom comes from a large scale hydrodynamic back-flow—with little or no charge in the fluid density. The region between $|\boldsymbol{k}| = 1$ Å^{-1} and $|\boldsymbol{k}| = 1.8$ Å^{-1} must be interpreted as a rather complicated transition region in which we pass from a high frequency phonon mode at 1 Å^{-1} to a single particle propagation at 2 Å^{-1}.

Some crude considerations suggest the modified mass to be about $2{\cdot}3\ m_{He}$. In this case the difference, at the minimum of the energy curve, between experimental and theoretical values is about 5%.

This view of the roton excitation can be confirmed by the results of Bogoliubov (1947) and Feynman and Cohen (1956). In the first paper the spectrum of elementary excitations becomes, for large p, the spectrum of a free particle. In the second paper, the authors describe the energy spectrum of the impurity atom as a spectrum of a free particle with an effective mass greater than the mass of atom. We refer also to a paper by Miller, Pines and Nozières (1962): "The Feynman quasi-particles for values of k lying between 1 Å^{-1} and 2 Å^{-1} behave rather like slow-moving impurity atoms in their coupling to the long wavelength excitations."

We should like to notice that "verbally" the last treatment of roton is closer to the ideas of Tisza than those of Landau: Tisza (1947) wrote when comparing the two variants of the two-fluid theory: "In Landau's theory the excited atoms of the Bose–Einstein theory are replaced by 'rotons'." Of course the wave function (7) proposed by Tisza has no chance to give a reasonable energy spectrum for helium.

Finally, Lee and Mohling (1959) examined experimental data for the total cross-section for the inelastic scattering of cold neutrons in liquid He II. They concluded that the projection of the angular momentum on the direction of the momentum $\boldsymbol{p}$ of the roton is equal to zero. This means that the roton excitations have zero angular momentum and confirms the conclusions that they have lost their rotational character.

Landau's two-fluid theory is still accepted as the best theory explaining most of the events in helium II. Again in summary, in this model the liquid was considered as consisting of two components, the normal and superfluid ones. The superfluid can undergo only irrotational, potential flow. The frictional force (mutual friction) between two components is absent. The superfluid component flows also without any interaction with solid boundaries. Such a picture of this component is contained in the condition that curl $\boldsymbol{v}_s \equiv 0$ in the liquid as a whole.

The disagreement between the theory and the results of experiments on both the free surface of rotating helium and on the critical velocity in helium showed that one needs an additional idea. Some suggestions followed from the fact that the observed rotation of the superfluid component was in contradiction to the assumption that curl $\boldsymbol{v}_s \equiv 0$.

The idea is inherent in a hypothesis pronounced long ago about the existence of vortex sheets (H. London, 1946; Onsager, 1948) and especially about the vortex lines and quantization of the superfluid circulation (Onsager, 1948) in helium. This last idea was essentially developed by Feynman (1955). At the beginning he considered the wave function of the current carrying state of helium at absolute zero. The wave function is assumed to be of the form

$$\psi_{\text{flow}} = [\exp i \sum_j s(\boldsymbol{r}_j)]\Phi \tag{24}$$

where the phase s is a function varying very slightly by displacements over distances of the order of the atomic spacing. The local velocity is defined by

$$\boldsymbol{v}_s = \frac{\hbar}{m} \nabla s. \tag{25}$$

From this definition it follows that the motion is irrotational, because curl $\boldsymbol{v}_s = 0$. If the region is simply connected and the walls fixed, the only solution of (25) is $\boldsymbol{v}_s = 0$. In this case it is clear that the circulation $\oint (\boldsymbol{v}_s \cdot d\boldsymbol{r})$ vanishes. A quite different situation is seen if the region is multiply connected even when the walls are fixed.

An example of a multiply connected region is a volume with a cylindrical hole of small diameter running through the liquid. In this case there are non-vanishing circulations around every circuit enclosing the hole. The values of the circulation are finite numbers. The symmetry requirements for ψ_{flow} (Feynman, 1955) make it possible to find these numbers. Let us consider a long closed chain of atoms changing their positions in such a way that

$r_j \to r_j + \Delta r_j = r_{j+1}$; for example, every atom is displaced from its initial position to the position of its nearest neighbour on the right-hand side. Such a movement is a special kind of permutation of Bose atoms under which $|\psi(1,\ldots,N)|^2$ cannot be changed. However, the change of phase of (24) is

$$\sum_{\text{ring}} (\text{grad } s(\boldsymbol{r}_i) \cdot \Delta \boldsymbol{r}_i) \to \int \frac{m}{\hbar} (\boldsymbol{v}_s(\boldsymbol{r}) \cdot d\boldsymbol{r}). \tag{26}$$

The states will be indistinguishable if the change of phase is equal to $n \cdot 2\pi$, where n is an integer. This condition leads to the quantization of circulation, namely

$$\oint (\boldsymbol{v}_s \cdot d\boldsymbol{r}) = \frac{n}{m} 2\pi\hbar = \frac{n}{m} h = \oiint (\text{curl } \boldsymbol{v}_s \cdot d^2\boldsymbol{S}). \tag{27}$$

For a simply connected region the circulation must vanish ($n = 0$) around any circuit. In a multiply connected region the irrotation condition is fulfilled through most of the superfluid. Only the circulations around holes are different from zero. Hence Feynman assumes the existence of holes in the superfluid component of rotating helium. They resemble a free vortex known from the hydrodynamics of an ordinary fluid. Feynman assumes that the hole radius a, at the centre, is of the order of the atomic spacing and he took arbitrarily $a = 4{\cdot}0$ Å. Some later estimations gave $a \sim 0{\cdot}5$ Å (Vinen, 1961a). Hence in the superfluid component there are no real macroscopic holes and one can speak only about vortex lines.

In the vortex lines there is contained some amount of energy. The kinetic energy per unit length is given by

$$\varepsilon = \frac{\pi \varrho \hbar^2}{m^2} \ln \frac{b}{a} \tag{28}$$

where b is an appropriate radius, of the order of the spacing between the vortex lines.

Approximate estimates performed by Feynman show that for an angular velocity of rotation $\omega \sim 1$ radian/sec the vortex lines are about 0·02 cm apart.

Now Landau's picture of helium II is changed. The condition curl $\boldsymbol{v}_s = 0$ is fulfilled over almost the whole of the liquid except around the vortex lines where curl $\boldsymbol{v}_s$ is infinite. Due to the presence of the vortex lines rotating helium behaves like an ordinary liquid.

A further very fruitful assumption of Feynman reads that "the contact between flowing helium and walls is a possible source of vortex lines". For the creation of vortex lines in the flowing fluid a finite energy is needed. The flow velocity below which the creation of vortex line is energetically disadvantageous is identified by Feynman with the critical velocity v_c. Therefore it is reasonable to assume that the creation of vortex lines is responsible for the frictional forces observed above the critical velocity. Following this assumption Feynman estimated the magnitude to be $v_c \sim 100$ cm/sec. This result is in good agreement with the experimental value $v_c \sim 70$ cm/sec. We remember that the considerations following from the Landau spectrum gave $v_c \sim 8000$–$24{,}000$ cm/sec.

It is convenient to treat the energy ε and momentum p of vortex lines as the energy and momentum of localized excitations (Atkins, 1959; Geilikman, 1959) and use directly Landau's definition of the critical velocity $v_c = (\varepsilon(p)/p)_{\min}$ given by (5). For a channel of width d we have (Lamb, 1952)

$$\varepsilon = \varrho \frac{\hbar^2}{2m^2}\left(\ln \frac{8d}{a} - \frac{7}{4}\right),$$

$$p = \pi \varrho \frac{\hbar}{m} d^2. \tag{29}$$

We must stress that vortex lines treated as excitations belong to the superfluid component. The collisions between lines and rotons or phonons give rise to the mutual friction between the two components which was not considered in Landau's theory.

The experimental evidence of the existence of vortex lines was given first by Hall and Vinen (1956a). They observed an extra attenuation of second sound in uniformly rotating helium. This fact was explained by a mutual friction of the two components. In their next paper (1956b) the authors measured the coefficients B, B' which appear in the expression for the total mutual friction F_{sn} (per unit volume) acting on the normal component. The coefficients were calculated from a new theory describing roton–vortex line collisions.

> The success of the theory given in this paper in accounting for the existence, the magnitude and the temperature-dependence of the mutual friction that is observed in rotating helium II is evidence that the theory is essentially correct, and, of course, that Feynman's theory of rotating helium, on which it is based, is also correct. More precisely, the evidence is in favour of a model of rotating helium in which the superfluid vorticity is confined to a uniform array of vortex lines, the number of lines per unit area perpendicular to the axis of rotation being approximately $2\omega m/\hbar$; in view of the remarks made in § 2, we cannot assert from the experimental results, as Feynman does on theoretical grounds, that the circulation round each line is strictly quantized.

The results of Hall and Vinen (1956b) suggested that the attenuation of second sound observed in their earlier paper (Hall and Vinen, 1956a) can be explained by the presence of vortex lines. Direct experimental evidence that the superfluid circulation is quantized was given finally by Vinen (1961b), Whitmore and Zimmerman (1965), and Rayfield and Reif (1963, 1964).

We must emphasize that although the existence of the quantized vortex lines is established beyond all doubt, as yet it is not clear in what way these lines are created. From the classical point of view a transition from a state of potential flow ($v < v_c$) to a state with vorticity ($v > v_c$) is forbidden. Hence such a transition must be of quantum mechanical origin.

In the previously mentioned Bogoliubov theory of a "nearly

perfect" Bose gas the fundamental importance of a "degenerate condensate" for superfluidity was emphasized. A further development of this idea was given by Onsager and Penrose (1956) and by Bogoliubov (1961, 1963).

The idea of a condensate originates from the property of an ideal Bose gas. In an ideal Bose gas below a definite temperature the zero momentum state is occupied by a macroscopically large occupation number of particles. The occupation numbers of the other states are vanishingly small. This can be expressed as follows:

$$\lim_{\substack{V \to \infty \\ N/V = \text{const}}} \frac{n_0}{N} = \alpha, \qquad \lim_{\substack{V \to \infty \\ N/V = \text{const}}} \frac{n_p}{N} = 0, \quad p \neq 0. \tag{30}$$

n_p is occupation number of the state with momentum $\boldsymbol{p}$, n_0 is the number of particles in the condensate, N the total number of particles. In a perfect gas $\alpha \to 1$ as $\theta \to 0$. For strongly interacting particles n_0 is a finite fraction of the total number of particles, even at $\theta = 0$ but does not tend to unity. For helium II, e.g. $\alpha \sim 0{\cdot}08$ (Onsager and Penrose, 1956). Onsager and Penrose proved that Bose–Einstein condensation takes place in liquid helium at absolute zero. They proved also that for a system with a condensate the mean value $\langle \psi(t, r) \rangle \neq 0$ where $\psi(t, r)$ is the Bose operator (see also Bogoliubov, 1947). This property of the system with condensate allows us to obtain from the microscopic theory the hydrodynamical equations, with two velocities, describing helium II. This was done by Bogoliubov (1963) and independently by Hohenberg and Martin (1965). Such a derivation of the hydrodynamic equations enables us to calculate from the microscopic theory the spectrum of the elementary excitations for small momenta. In the brief introduction to the method of the derivation of the hydrodynamic equations we shall follow Bogoliubov (1963).

Consider a system of Bose particles with the Hamiltonian

$$\hat{H}[\eta, U] = \hat{H} + \delta\hat{H}_t[\eta, U] \tag{31}$$

where

$$\hat{H} = \frac{1}{2m}\int \nabla\psi^+(t, \boldsymbol{r})\nabla\psi(t, \boldsymbol{r})\, d^3\boldsymbol{r} - \lambda \int \psi^+(t, \boldsymbol{r})\psi(t, \boldsymbol{r})\, d^3\boldsymbol{r}$$

$$+\frac{1}{2}\int V(\boldsymbol{r}-\boldsymbol{r}')\psi^+(t, \boldsymbol{r})\psi^+(t, \boldsymbol{r}')\psi(t, \boldsymbol{r}')\psi(t, \boldsymbol{r})\, d^3\boldsymbol{r}'\, d^3\boldsymbol{r},$$

$$\delta\hat{H}_t(\eta, U) = \int \{\delta\eta(t, \boldsymbol{r})\psi^+(t, \boldsymbol{r}) + \delta\eta^+(t, \boldsymbol{r})\psi(t, \boldsymbol{r})\}\, d^3\boldsymbol{r} \tag{32}$$

$$+\int \delta U(t, \boldsymbol{r})\hat{\varrho}(t, \boldsymbol{r})\, d^3\boldsymbol{r},$$

$$\hat{\varrho}(t, \boldsymbol{r}) = \psi^+(t, \boldsymbol{r})\psi(t, \boldsymbol{r}).$$

Here, $\hbar = 1$, and λ is a constant. An external scalar potential δU and "sources of particles" $\delta\eta$, $\delta\eta^*$ are introduced into the Hamiltonian. They are given c-functions. We assume that our system is at $t = -\infty$ in thermodynamic equilibrium. After the adiabatic switching on of a time-dependent, small perturbation δH_t we shall have a small deviation of the system from thermodynamic equilibrium.

In the Hamiltonian (31) there appear terms linear in the operators ψ, ψ^+. The reason of this is that we shall use explicitly the fact that for a system with a condensate $\langle\psi\rangle \neq 0$. Therefore the Hamiltonian $\hat{H}$ (we must distinguish between $\hat{H}$ and $\hat{H}[\quad]$!) has more "symmetry" than our description of the system. Therefore the "symmetry" of the Hamiltonian $\hat{H}$, in this special case gauge invariance, must be "broken" by the addition of terms linear in ψ, ψ^+ included in $\delta\hat{H}_t$ (see, for example, Wagner, 1966; Stern, 1966).

It was shown by Bogoliubov (1963) that if some kind of symmetry (translational, rotational, gauge invariance) is broken then in statistical physics we are dealing not with normal averages but with so called "quasi-averages". For example the normal average $\langle\psi\rangle$ always vanishes, whereas the appropriate quasi-average is different from zero. The symmetries of the system are connected with conservation laws. For example, the translational, rotational and gauge invariance symmetries are connected with the conservation laws of momentum, angular momentum, and number of particles respectively.

If the system is slightly disturbed from thermodynamic equilibrium it can be described by means of the hydrodynamic equations. They can be obtained after a calculation of the time derivatives of the following local quantities:

$$\varrho(t, \boldsymbol{r}) = \langle\psi^+(t, \boldsymbol{r})\psi(t, \boldsymbol{r})\rangle,$$

$$\boldsymbol{j}(t, \boldsymbol{r}) = \frac{i}{2}\langle(\nabla\psi^+(t, \boldsymbol{r}))\psi(t, \boldsymbol{r}) - \psi^+(t, \boldsymbol{r})(\nabla\psi(t, \boldsymbol{r}))\rangle,$$

$$\begin{aligned}\varrho(t, \boldsymbol{r})\varepsilon(t, \boldsymbol{r}) = {} & -\frac{1}{4m}\langle(\nabla^2\psi^+(t, \boldsymbol{r}))\psi(t, \boldsymbol{r}) \\ & +\psi^+(t, \boldsymbol{r})(\nabla^2\psi(t, \boldsymbol{r}))\rangle \\ & +\frac{1}{2}\int V(\boldsymbol{r}-\boldsymbol{r}')\langle\psi^+(t, \boldsymbol{r})\psi^+(t, \boldsymbol{r}')\psi(t, \boldsymbol{r}')\psi(t, \boldsymbol{r})\rangle\, d^3\boldsymbol{r}',\end{aligned} \tag{33}$$

$$\Phi(t, \boldsymbol{r}) = \langle\psi(t, \boldsymbol{r})\rangle \equiv a e^{i\chi},$$

$$\langle\hat{B}\rangle = \mathrm{Tr}(D_0\hat{B}), \quad D_0 = D(-\infty) = \frac{e^{-\hat{H}/\theta}}{\mathrm{Tr}(e^{-\hat{H}/\theta})}.$$

$\varrho\varepsilon$ is the mean energy per particle, $\psi(t, \boldsymbol{r})$ and $\psi^+(t, \boldsymbol{r})$ are operators in the Heisenberg representation with the total Hamiltonian $\hat{H}[\eta, U]$; θ is now in energy units.

In the superfluid system there can exist two independent motions. Motion of the superfluid component occurs with a velocity

$$\boldsymbol{v}_s = \frac{1}{m}\nabla\chi \tag{34}$$

where χ is the phase of $\langle\psi\rangle$; the motion of the normal component is described by the velocity $\boldsymbol{v}_n$ defined by the formula

$$\boldsymbol{j} = m\varrho_s\boldsymbol{v}_s + m\varrho_n\boldsymbol{v}_n \tag{35}$$

Here $\boldsymbol{j}$ is defined by (33), $\boldsymbol{v}_s$ by (34),

$$\varrho_s = \frac{\varrho}{m}\frac{\partial F}{\partial u} \quad \text{where} \quad u = v_s^2/2,$$

$$F = -\frac{1}{N}\,\theta \ln \mathrm{Tr}\{e^{-H/\theta}\} \quad \text{is free energy per particle,}$$

ϱ is defined by (33) and $\varrho_n = \varrho - \varrho_s$.

The time derivatives $\partial\varrho/\partial t$, $\partial \boldsymbol{j}/\partial t$, $\partial\varrho\varepsilon/\partial t$, $\partial\langle\psi\rangle/\partial t$ can be obtained after calculating the commutators of suitable operators with the Hamiltonian $\hat{H}[\eta, U]$. Therefore in the expressions for the commutators terms proportional to $\delta\eta$, $\delta\eta^*$, δU will be present. From the expressions for $\partial\langle\psi\rangle/\partial t$, $\partial\langle\psi^+\rangle/\partial t$ it is possible to calculate $\partial \boldsymbol{v}_s/\partial t$.

The function $\Phi = \langle\psi\rangle$ is treated as a wave function of the condensate. By means of this function it is possible to obtain on expression for the current of the condensate $\boldsymbol{j}_c$

$$\boldsymbol{j}_c = \tfrac{1}{2}i(\Phi\nabla\Phi^* - \Phi^*\nabla\Phi) = a^2\nabla\chi = \varrho_c m \boldsymbol{v}_s \tag{36}$$

$\varrho_c = a^2$ is the density of the condensate.

For small deviations from thermodynamic equilibrium we can put

$$\Phi = \Phi_0 + \delta\Phi = \sqrt{\varrho_0} + \delta\Phi, \quad \Phi^* = \Phi_0 + \delta\Phi^* = \sqrt{\varrho_0} + \delta\Phi \tag{37}$$

where ϱ_0 is the density of the condensate for thermodynamic equilibrium. In this case formula (36) gives

$$\delta \boldsymbol{j}_c = \tfrac{1}{2}i\sqrt{\varrho_0}(\nabla\delta\langle\psi^+\rangle - \nabla\delta\langle\psi\rangle) = \varrho_0 m\, \delta \boldsymbol{v}_s \tag{38}$$

where $\delta\langle\psi\rangle = \delta\Phi$.

In the expressions for $\partial\varrho/\partial t$, $\partial \boldsymbol{j}/\partial t$, $\partial \boldsymbol{v}_s/\partial t$, there appear mean values of the form

$$B = \langle(D_1\psi^+(t, \boldsymbol{r}))\,(D_2\psi^+(t, \boldsymbol{r}'))\,(D_3\psi(t, \boldsymbol{r}'))\,(D_4\psi(t, \boldsymbol{r}))\rangle \tag{39}$$

where D_j are integers or differential operators. In the case of thermodynamic equilibrium these mean values are functions of some parameters which are constant in space and time. In the case of an ordinary liquid it is convenient to choose for these parameters ϱ, θ, $v^{(\alpha)}$: the particle density, temperature, and drift velocity. The superfluid is characterized by two independent motions, therefore we need more parameters for which we choose ϱ, θ, $v_s^{(\alpha)}$, $v_n^{(\alpha)}$.

In order to obtain hydrodynamic equations describing non-equilibrium processes we must assume that all the mean values (39) change very slowly in space and time. They are now functions of the local quantities $\varrho(t, \boldsymbol{r})$, $\theta(t, \boldsymbol{r})$, $v_s^{(\alpha)}(t, \boldsymbol{r})$, $v_n^{(\alpha)}(t, \boldsymbol{r})$ and their spatial derivatives. Because we are interested in non-equilibrium states very close to the equilibrium state the local parameters must vary really slowly in space and time. In consequence the gradients of ϱ, θ, $v_s^{(\alpha)}$, $v_n^{(\alpha)}$ must be small and the mean values (39) can be written in the form

$$\begin{aligned} B(t, \boldsymbol{r}, \boldsymbol{R}) = {} & B^{(0)}(\varrho(t, \boldsymbol{r}), \theta(t, \boldsymbol{r}), v_s^{(\alpha)}(t, \boldsymbol{r}), v_n^{(\alpha)}(t, \boldsymbol{r}); \boldsymbol{R}) \\ & + B^{(1)}(t, \boldsymbol{r}, \boldsymbol{R}) + B^{(2)}(t, \boldsymbol{r}, \boldsymbol{R}) \dots \qquad (40) \\ \boldsymbol{R} = {} & \boldsymbol{r} - \boldsymbol{r}'. \end{aligned}$$

In (40) $B^{(0)}$ depends only on the local variables. It is obtained by means of exchanging in the functional dependence of the equilibrium mean value B the constant parameters by suitable local ones. The function $B^{(1)}$ is a form linear in the gradients of ϱ, θ, $v_s^{(\alpha)}$, $v_n^{(\alpha)}$, etc. We remark that the expansion of mean values in terms of gradients is fundamental for the derivation of hydrodynamic equations from the Boltzman equation (see, for example, Chapman and Cowling, 1952).

If in the mean value (40) only the first term $B^{(0)}$ is taken into account, we obtain the hydrodynamic equations of the "ideal" Bose liquid, without consideration of the viscosity of the normal component. Such equations were obtained first by Landau (1941) and with dissipative terms by Khalatnikov (1952a, reprinted in Part 2 as paper T6 on p. 314).

Recently Bogoliubov (1963) has obtained the Landau equations from the microscopic theory (see also Hohenberg and Martin, 1965) by considering an external scalar potential and sources of particles. He showed that all mean values like $B^{(0)}$ can be eliminated from the equations for $\partial\varrho/\partial t$, $\partial \boldsymbol{j}/\partial t$, $\partial\varrho\varepsilon/\partial t$, $\partial\langle\psi\rangle/\partial T$. He obtained, for example for $v_s^2 = 0$ (see also Martin and Schwinger, 1959).

$$-P(\varrho, \theta) = -\frac{1}{m}\left\langle \frac{\partial\psi^+}{\partial r_\alpha} \frac{\partial\psi}{\partial r_\alpha} \right\rangle + \frac{1}{2}\int \frac{\partial V(\boldsymbol{R})}{\partial R_\alpha} R_\alpha \langle\psi^+\psi^+\psi\psi\rangle \, d^3\boldsymbol{R}, \tag{41}$$

or (see also Hohenberg and Martin, 1965)

$$\frac{1}{a^2}\int V(\boldsymbol{R})\langle\psi^+(t, \boldsymbol{r}')\psi(t, \boldsymbol{r}')\psi(t, \boldsymbol{r})\rangle\Phi^*(t, \boldsymbol{r})\, d^3\boldsymbol{R} = \Lambda(\varrho, \theta, u) - \frac{mv_s^2}{2}, \quad u = \frac{(\boldsymbol{v}_s - \boldsymbol{v}_n)^2}{2}. \tag{42}$$

Λ is chemical potential per particle (in Landau's equations he used the chemical potential $\mu = \Lambda/m$). Finally, the hydrodynamical equations of the two-fluid theory can be written in the form

$$\frac{\partial\varrho}{\partial t} + (\nabla\cdot[\varrho_s\boldsymbol{v}_s + \varrho_n\boldsymbol{v}_n]) = ia(\zeta^* - \zeta), \tag{43}$$

$$m\frac{\partial}{\partial t}(\varrho_s v_s^{(\alpha)} + \varrho_n v_n^{(\alpha)}) + m\sum_\beta \frac{\partial}{\partial r_\beta}(\varrho_s v_s^{(\alpha)} v_s^{(\beta)} + \varrho_n v_n^{(\alpha)} v_n^{(\beta)}) + \frac{\partial P}{\partial r_\alpha} = -\varrho\frac{\partial\delta U}{\partial r_\alpha} + ima(\zeta^* - \zeta)v_s^{(\alpha)},$$
$$\zeta = \delta\eta(t, \boldsymbol{r})e^{-i\chi}, \tag{44}$$

$$m\frac{\partial\boldsymbol{v}_s}{\partial t} + \nabla\left(m\left[(\boldsymbol{v}_s\cdot\boldsymbol{v}_n) - \frac{v_n^2}{2}\right] + \Lambda\right) = -\nabla\,\delta U, \tag{45}$$

$$\frac{\partial(\varrho S)}{\partial t} + (\nabla\cdot(\varrho S\boldsymbol{v}_n)) = 0. \tag{46}$$

Here S is the entropy per particle and P the pressure.

In order to obtain (46) we have used the following formulae:

$$\varrho\varepsilon(\varrho, \theta, v_s^{(\alpha)}, v_n^{(\alpha)}) = \varrho E(\varrho, \theta, u) + \frac{m\varrho v_n^2}{2}$$
$$+ m\varrho_s(\boldsymbol{v}_s - \boldsymbol{v}_n \cdot \boldsymbol{v}_n),$$
$$u = \frac{(\boldsymbol{v}_s - \boldsymbol{v}_n)^2}{2}, \quad E = F - \theta\frac{\partial E}{\partial \theta} \tag{47}$$
$$P = \varrho^2 \frac{\partial F}{\partial \varrho}, \qquad \Lambda = F + \varrho\frac{\partial F}{\partial \varrho}$$
$$S = -\frac{\partial F}{\partial \theta}.$$

The hydrodynamic equations with viscous terms were derived in the Bogoliubov scheme by Galasiewicz (1964, 1967) and Krasnikov (1967). In the term $B^{(1)}$ which is linear in the gradients there appear many coefficients which one assumed to be given separately. According to the additional considerations of the possibility of the existence of thermodynamic equilibrium if $\delta\eta = \delta\eta(\boldsymbol{r})$, $\delta U = \delta U(\boldsymbol{r})$ (time independent!) it was possible to find some connections between the coefficients introduced. The independent ones can be identified with the kinetic coefficients of the second viscosity $\zeta_1 = \zeta_4$, ζ_2, ζ_3, the coefficient of thermal conductivity $\varkappa$, and the coefficient of first viscosity η appearing in the equations given by Khalatnikov (1952a).

The assumption about the phonon–roton spectrum of the elementary excitations in helium and then the calculation of effective differential cross-sections for the scattering of quasi-particles by each other allowed Landau and Khalatnikov (1949) to find the temperature dependence of η and Khalatnikov (1962b) to find that of $\varkappa$ and η.

Khalatnikov assumed that the distribution function for elementary excitations in helium satisfies a Boltzman equation. From this equation Khalatnikov (1952a) derived the hydrodynamical equations for helium II with viscous terms. He obtained also the genera expressions for the coefficients of second viscosity and proved the Onsager relation $\zeta_1 = \zeta_4$.

The hydrodynamic equations (43)–(46) are non-linear. If we are interested in infinitesimal deviations from thermodynamic equilibrium, the unknown functions ϱ, θ, $v_s^{(\alpha)}$, $v_n^{(\alpha)}$ can be written in the form

$$\varrho(t, \boldsymbol{r}) = \varrho^0 + \delta\varrho(t, \boldsymbol{r}) \tag{48}$$

where ϱ^0, θ^0, v_s^0, v_n^0 are constant parameters describing thermodynamic equilibrium. We are interested in a special kind of equilibrium when $\boldsymbol{v}_s^0 = \boldsymbol{v}_n^0 = 0$. For the variations $\delta\varrho$, $\delta\theta$, $\delta\boldsymbol{v}_s$, $\boldsymbol{v}_n$ we have now the linearized hydrodynamic equations

$$\frac{\partial \delta\varrho}{\partial t} + \varrho_s \operatorname{div} \delta\boldsymbol{v}_s + \varrho_n \operatorname{div} \delta\boldsymbol{v}_n = i\sqrt{\varrho_0}\,(\delta\eta^* - \delta\eta), \tag{49}$$

$$m\varrho_s \frac{\partial \delta\boldsymbol{v}_s}{\partial t} + m\varrho_n \frac{\partial \delta\boldsymbol{v}_n}{\partial t} = -\nabla\delta P - \varrho\nabla\delta U, \tag{50}$$

$$m\frac{\partial \delta\boldsymbol{v}_s}{\partial t} = -\nabla\delta\Lambda - \nabla\delta U, \tag{51}$$

$$\varrho\frac{\partial \delta S}{\partial t} + S\frac{\partial \delta\varrho}{\partial t} + \varrho S \operatorname{div} \delta\boldsymbol{v}_n = 0 \tag{52}$$

where

$$\sqrt{\varrho_0} = \langle\psi\rangle = \langle\psi^+\rangle,$$
$$\delta\Lambda = \frac{\partial\Lambda}{\partial\delta}\,\delta\varrho + \frac{\partial\Lambda}{\partial\theta}\,\delta\theta = -S\delta\theta + \frac{1}{\varrho}\,\delta P. \tag{53}$$

Now we shall put for simplicity $\delta U = 0$.

In order to switch δH_t on adiabatically we take $\delta\eta$ in the form

$$\begin{aligned}\delta\eta(t, \boldsymbol{r}) &= e^{-i\omega t + \varepsilon t + i(\boldsymbol{k}\cdot\boldsymbol{r})}\,\delta\eta(k)\\ &+ e^{i\omega t + \varepsilon t - i(\boldsymbol{k}\cdot\boldsymbol{r})}\,\delta\eta(-k), \quad \varepsilon > 0, \quad \varepsilon \to 0.\end{aligned} \tag{54}$$

The presence of $\varepsilon > 0$ preserves the vanishing of $\delta\eta$ for $t = -\infty$. The solutions of (49)–(52) are linear functions of $\delta\eta$, $\delta\eta^*$; therefore they can be taken also in the form (54). For the Fourier components we have now the following equations:

$$\varrho_s\delta X(k)+\varrho_n\delta Y(k)-E\delta\varrho(k) = \sqrt{\varrho_0}(\delta\eta^*_{-k}-\delta\eta_k), \tag{55}$$

$$Em\varrho_s\delta X(k)+Em\varrho_n\delta Y(k)-k^2\left(\frac{\partial P}{\partial\varrho}\right)_\theta\delta\varrho(k)$$
$$-k^2\left(\frac{\partial P}{\partial\theta}\right)_\varrho\delta\theta(k) = 0, \tag{56}$$

$$Em\varrho\delta X(k)-k^2\left(\frac{\partial P}{\partial\varrho}\right)_\theta\delta\varrho(k)-k^2\left[\left(\frac{\partial P}{\partial\theta}\right)_\varrho-\varrho S\right]\delta\theta(k) = 0, \tag{57}$$

$$-\varrho S\delta Y(k)+E\left[\varrho\left(\frac{\partial S}{\partial\varrho}\right)_\theta+S\right]\delta\varrho(k)$$
$$+E\varrho\left(\frac{\partial S}{\partial\theta}\right)_\varrho\delta\theta(k) = 0, \tag{58}$$

$$E = \omega+i\varepsilon, \quad \varepsilon > 0, \quad \varepsilon \to 0$$
$$\delta X(k) = (\boldsymbol{k}\cdot\delta\boldsymbol{v}_s(k)), \quad \delta Y(k) = (\boldsymbol{k}\cdot\delta\boldsymbol{v}_n(k)) \tag{59}$$

From (55)–(58) we have

$$(\boldsymbol{k}\cdot\delta\boldsymbol{v}_s(k)) = \frac{m(c_1k)^2}{m\varrho_sM_1M_2}\left\{(c_2k)^2\right.$$
$$\left.-\frac{E^2}{\varrho}\left[\varrho_s+\left(\frac{c_2}{c_1}\right)\varrho_n\right]\right\}\sqrt{\varrho_0}\,(\delta\eta^*_{-k}-\delta\eta_k), \tag{60}$$

$$(\boldsymbol{k}\cdot\delta\boldsymbol{j}(k)) = m(\varrho_s\delta X+\varrho_n\delta Y)$$
$$= -\frac{m(c_1k)^2}{M_1}\sqrt{\varrho_0}(\delta\eta^*_{-k}-\delta\eta_k), \tag{61}$$

$$\delta\varrho(k) = -\frac{mE}{M_1}\sqrt{\varrho_0}(\delta\eta^*_{-k}-\delta\eta_k), \tag{62}$$

where

$$c_1^2 \approx \frac{1}{m}\left(\frac{\partial P}{\partial\varrho}\right)_s \approx \frac{1}{m}\left(\frac{\partial P}{\partial\varrho}\right)_\theta, \quad c_2^2 \approx \frac{S^2\varrho_s\theta}{m\varrho_nc_v}, \tag{63}$$

$$c_v = \theta\left(\frac{\partial S}{\partial\theta}\right)_\varrho,$$
$$M_i = (E^2-c_i^2k^2), \qquad i = 1, 2\ldots$$

$c_{1,2}$ are the first and second sound velocities, respectively.

When deriving the hydrodynamic equations, we have emphasized the smallness of the deviation from thermodynamic equilibrium. This deviation leads to a local equilibrium described by the local quantities $\varrho(t, r)$, $\theta(t, r)$ $v_s(t, r)$, $v_n(t, r)$. The local equilibrium is a consequence of the frequent collisions between the particles. Therefore, the hydrodynamic region is a collision-dominated one. In such a region the conventional methods for computing the Green functions, giving us the spectrum of the elementary excitations, cannot be successful because they are based on expansions in terms of certain parameters describing the number of collisions in the system. On the other hand, the description of the system in the collision-dominated region both in terms of Green functions as well as in terms of the hydrodynamical equations must be equivalent. In order to have the mathematical description of this situation it is convenient to use the formula connecting the variations of the mean values of operators with temperature-dependent Green functions (e.g. Bogoliubov, 1963; Zubarev, 1960) or correlation functions (Kadanoff and Martin, 1963; Hohenberg and Martin, 1965). We shall now give a brief derivation of this important formula.

The mean value $\langle \hat{B} \rangle$ of an operator $\hat{B}$ is equal to $\langle \hat{B} \rangle = \mathrm{Tr}\,\langle \hat{D}\hat{B} \rangle$ where $\hat{D}$ is the statistical operator. At $t = -\infty$ we have thermodynamical equilibrium and $\hat{D} = \hat{D}_0$ given by (33). After the time-dependent, small perturbation δH_t is switched on the operator $\hat{D}_0$ will change to $\hat{D}_S(t) = \hat{D}_0 + \delta\hat{D}_S(t)$. We can find $\delta\hat{D}_S(t)$ (in the Schrödinger representation) from the equation

$$i \frac{d\hat{D}_s(t)}{dt} = [\hat{H}[\eta, U], \hat{D}_s(t)] \tag{64}$$

it is equal to

$$\delta\hat{D}_s(t) = \frac{1}{i} \int_{-\infty}^{t} e^{\hat{H}(t-\tau)} [\delta\hat{H}_\tau, \hat{D}_0]\, e^{-i\hat{H}(t-\tau)}\, d\tau. \tag{65}$$

Putting

$$\delta\langle \hat{B} \rangle = \langle \hat{B} \rangle - \mathrm{Tr}\,\{\hat{D}_0\hat{B}\} \tag{66}$$

we obtain for an operator $\hat{B}$ (which does not depend explicitly on the time)

$$\delta\langle\hat{B}\rangle = -i\int_{-\infty}^{t}\frac{\mathrm{Tr}\,\{e^{-H/\theta}[\hat{B}(t),\,\delta\hat{H}_\tau(\tau)]\}}{\mathrm{Tr}\,e^{-H/\theta}}\,d\tau$$

$$= \int_{-\infty}^{+\infty} \ll \hat{B}(t);\quad \delta\hat{H}_\tau(\tau) \gg^r d\tau. \tag{67}$$

In (67) $\ll;\gg^r$ indicates a retarded Green function defined by

$$\ll \hat{A}(t);\,\hat{B}(\tau)\gg^r \;=\; G_r(t-\tau) = -i\theta(t-\tau)\langle[\hat{A}(t);\,\hat{B}(\tau)]\rangle$$

$$\theta(t) = \begin{cases} 1, & t>0 \\ 0, & t<0. \end{cases} \tag{68}$$

$\hat{A}(t)$, $\hat{B}(\tau)$ are operators in the Heisenberg representation with the Hamiltonian $\hat{H}$.

If $\delta\hat{H}_t$ is defined by (32) and (54) formula (67) has in the momentum representation the form

$$\delta B(k) = 2\pi\{\ll \hat{B}(k);\; a_k \gg^r_E \delta\eta^*_{-k} + \ll \hat{B}(k);\; a^+_k \gg^r_E \delta\eta_k +$$
$$+ \ll \hat{B}(k);\; \hat{\varrho}_{-k} \gg^r_E \delta U(k). \tag{69}$$

The derivatives

$$\frac{\delta B(k)}{\delta\eta_k} = 2\pi \ll \hat{B}(k);\; a^+_k \gg^r_E,$$
$$\frac{\delta B(k)}{\delta U(k)} = 2\pi \ll \hat{B}(k);\; \hat{\varrho}_k \gg^r_E \tag{70}$$

can be expressed in terms of retarded Green functions. If we put $\hat{B} = \hat{\varrho}, \hat{j}$ it is possible to obtain formulae for the density and current variations. If we put $\hat{B} = \psi,\ \psi^+$ it is possible, because of formula (38), to obtain expressions for the variation of the superfluid velocity. On the other hand it is possible to calculate the derivatives (70) which interest us, independently from formulae (60)–(62).

A comparison of the results gives us expressions for the Green functions $\ll a_k;\, a_k \gg_E$, $\ll \hat{\varrho}_k;\, a_k \gg_E$ ... which was the basic idea of the calculations of Green functions in the "hydrodynamical" approximation. We must emphasize, however, that the description of a many body system in terms of Green functions and in terms of linearized hydrodynamic equations is the same only for small k and E, namely for $k \ll 1/l$, $E \ll 1/T$ where l is the mean free path and T the relaxation time.

From (70), (60), (38) we see that the one-particle Green functions $\ll a_k;\, a_k \gg_E$ have poles at the energies of the elementary excitations of two kinds (now we denote these energies by E) $E = c_1 k$, $E = c_2 k$, i.e. the energies of the first and second sound quanta (Bogoliubov, 1963).

From (62) we see that the Green functions $\ll \hat{\varrho}_k;\, a_k \gg_E$ describing the density oscillations have poles for the elementary excitations of one kind, namely $E = c_1 k$, i.e. energies of the first sound quanta (Galasiewicz, 1967).

Thus we have proved on the basis of the microscopic theory that the lowest excitations in a superfluid are of the phonon type.

In the case of the linearized hydrodynamical equations with viscous terms the poles of the Green functions will obtain imaginary parts describing the damping of first and second sound (Galasiewicz, 1964, 1967).

The examination of the single-particle Green functions for thermodynamic equilibrium ($E = 0$, no time dependence in (54)) gives

$$\ll a_k;\, a_{-k} \gg_0^r = \frac{m\varrho_0}{2\pi\varrho_s}\,\frac{1}{k^2}. \tag{71}$$

This is the "$1/k^2$ theorem" of Bogoliubov (1961) describing the special singularity of Green functions as $k \to 0$. This theorem plays an essential role in the examination of the long-range ordering in superfluid helium. On the basis of this theorem Kadanoff and Kane (1967) showed that Yang's criterion (Yang (1962)) for the existence of long-range order—ODLRO (standing for off-diagonal long-range order)—(criterion for superfluidity) is not satisfied in one and two dimensions.

The idea of the condensate and condensate wave function $\Phi = \langle\psi\rangle$ is important not only for the derivation of the complete system of hydrodynamic equations. With the help of this function it is possible also to consider the problem of vortices in superfluid helium (Vinen, 1966). The vortex lines are the lines along which the condensate wave function $\langle\psi\rangle$ vanishes (Iordansky, 1964). The circulation of a superfluid takes place around these lines. Thus the microscopic approach gives us the possibility for a uniform description of phenomena in helium II.

REFERENCES

ALLEN, J. F. PEIERLS, R. and UDDIN, ZAKI (1937) *Nature*, **140**, 62 (paper E5 on p. 92).

ALLEN, J. F. and JONES, H. (1938) *Nature*, **141**, 243 (paper E6 on p. 95).

ALLEN, J. F. MISENER, A. D. (1938) *Nature*, **141**, 75.

ANDRONIKASHVILI, E. (1946) *J. Phys. U.S.S.R.*, **10**, 201 (paper E11 on p. 154).

ATKINS, K. R. and OSBORNE, D. V. (1950) *Phil. Mag.* **41**, 1078.

ATKINS, K. R. (1959) *Liquid Helium*, Cambridge University Press, London.

BEAUMONT, C. F. A. and REEKIE, J. (1955) *Proc. Roy. Soc.* **A 228**, 363.

DE BOER J. (1963) *Liquid Helium* (Editor CARERI), Academic Press.

BOGOLIUBOV, N. N. (1947) *J. Phys. U.S.S.R.* **11**, 23 (paper T4 on p. 247).

BOGOLIUBOV, N. N. (1961) (2nd ed. 1963) *The Quasi-averages in Problems of Statistical Physics*, Dubna preprint (in Russian; English translation to be published by Gordon and Breach); see also (1960), *Physica* **26** (suppl.), 1.

BOGOLIUBOV, N. N. (1963) *On the Hydrodynamics of a Superfluid*, Dubna preprint (in Russian; English translation to be published by Gordon and Breach).

CHAPMAN, S. and COWLING, T. G. (1952) *The Mathematical Theory of Nonuniform Gases*, Cambridge University Press, London.

CHESTER, G. V. (1963) *Liquid Helium* (Editor CARERI), Academic Press.

DANA, L. J. and KAMERLINGH ONNES, H. (1926) *Leiden Comm.* Suppl. 179d.

DAUNT, J. G. and MENDELSSOHN, K. (1938a) *Nature* **141**, 911 (paper E9 on p. 111).

DAUNT, J. G. and MENDELSSOHN, K. (1938b) *Nature* **142**, 475.

DAUNT, J. G. and MENDELSSOHN, K. (1939a) *Nature* **143**, 719.

DAUNT, J. G. and MENDELSSOHN, K. (1939b) *Proc. Roy. Soc. (London)* A **170**, 423, 439.

EHRENFEST, P. (1933) *Leiden Comm.* Suppl. 75b, 8.

FEYNMAN, R. P. (1953) *Phys. Rev.* **91**, 1301.

FEYNMAN, R. P. (1954) *Phys. Rev.* **94**, 262.

FEYNMAN, R. P. (1955) *Prog. Low Temp. Phys.* (Editor GORTER), North-Holland, **1**, 17 (paper T5 on p. 268).

FEYNMAN, R. P. and COHEN, M. (1956) *Phys. Rev.* **102**, 1189.

FRENKEL, J. (1946) *Kinetic Theory of liquids*, Oxford University Press.

FRÖHLICH, H. (1937) *Physica* **4**, 639.

GALASIEWICZ, Z. M. (1964) *Assymptotical Formulae for Green's Functions in the Viscous Liquid Approximation for the Superfluid Systems of Bose-particles*, Dubna preprint (in Russian), revised and enlarged version (1970) *Superconductivity and Quantum Fluids*, Pergamon Press, Oxford.

GALASIEWICZ, Z. M. (1967) *Bull. de l'Acad. de Sci. Polon.* **15,** 191.

GEILIKMAN, B. T. (1959) *J.E.T.P.* **37,** 891; *Sov. Phys. — JETP* **10,** 635.

GORDON, W. L. SHAW, C. H. and DAUNT, J. G. (1958) *Phys. Chem. Solids* **5,** 117.

HALL, H. E. and VINEN, W. F. (1956a) *Proc. Roy. Soc.* A **238,** 204.

HALL, H. E. and VINEN, W. F. (1956b) *Proc. Roy. Soc.* A **238,** 215.

HAM, A. C. and JACKSON, L. C. (1954) *Phil. Mag.* **45,** 1084.

HENSHAW, D. G. and WOODS, A. D. B. (1961) *Phys. Rev.* **121,** 1266.

HOHENBERG, P. C. and MARTIN, P. C. (1965) *Ann. Phys. (N. Y.)* **34,** 291.

IORDANSKY, S. W. (1964) *Ann. Phys. (N. Y.)* **29,** 335.

KADANOFF, L. P. and MARTIN, P. C. (1963) *Ann. Phys. (N. Y.)* **24,** 419.

KADANOFF, L. P. and KANE, J. W. (1967) *Phys. Rev.* **155,** 80.

KAMERLINGH ONNES, H. (1911) *Proc. Roy. Acad., Amsterdam* **13,** 1093.

KAMERLINGH ONNES, H. (1922) *Trans. Faraday Soc.* **18,** No. 53.

KAMERLINGH ONNES, H. and BOKS, J. D. A. (1924) *Leiden Comm.*, **170b** (paper E1 on p. 59).

KAPITZA, P. (1938) *Nature* **141,** 74 (paper E7 on p. 99).

KAPITZA, P. (1941a) *J. Phys. U.S.S.R.* **4,** 181 (paper E10 on p. 114).

KAPITZA, P. (1941b) *J. Phys. U.S.S.R.* **5,** 59.

KEESOM, W. H. and WOLFKE, M. (1927) *Leiden Comm.* **190b,** 17 (paper E2 on p. 65).

KEESOM, W. H. and CLUSIUS, K. (1932) *Leiden Comm.* **219c** (paper E3 on p. 72).

KEESOM, W. H. and Miss KEESOM, A. P. (1936) *Physica* **3,** 359 (paper E4 on p. 89).

KEESOM, W. H. and MACWOOD J. E. (1938) *Physica* **5,** 737 (paper E8 on p. 103).

KEESOM, W. H. and TACONIS, K. (1937) *Physica* **4,** 28, 256.

KEESOM, W. H. and TACONIS, K. (1938) *Physica* **5,** 270.

KHALATNIKOV, I. M. (1952a) *J.E.T.P.* **23,** 8 (paper T6 on p. 314).

KHALATNIKOV, I. M. (1952b) *J.E.T.P.* **23,** 21. See also (1965) *An Introduction to the Theory of Superfluidity*, Benjamin, New York.

KHALATNIKOV, I. M. and CHERNIKOVA D. M. (1966) *Sov. Phys. J.E.T.P.* **23,** 274.

KIKOIN, A. K. and LASAREV, B. J. (1938) *Nature*, **141,** 912; 142, 289.

KRASNIKOV, W. A. (1967) Dissertation, Moscow State University.

KURTI, K., ROLLIN, B. V. and SIMON, F. (1936) *Physica* **3,** 266.

LAMB, H. (1952) Hydrodynamics, 6th ed., Cambridge University Press, London.

LANDAU, L. D. (1941) *J. Phys. U.S.S.R.* **5,** 71 (paper T1 on p. 191); *Collected Papers*, Pergamon Press, 1965, p. 301.

LANDAU, L. D. (1947) *J. Phys. U.S.S.R.* **11,** 91 (paper T3 on p. 243); *Collected Papers*, Pergamon Press, 1965, p. 466.

LANDAU, L. D. (1949) *Phys. Rev.* **75,** 884; *Collected Papers,* Pergamon Press, 1965, p. 474.

LANDAU, L. D. and KHALATNIKOV, I. M. (1949) *J.E.T.P.* **19,** 637, 709; *Collected Papers of L. D. Landau,* Pergamon Press, 1965, p. 494.

LANE, C. T. (1962) *Superfluid Physics,* McGraw-Hill, New York.

LEE, T. D. and MOHLING, F. (1959) *Phys. Rev. Letters* **2,** 284.

LIFSHITZ, E. M. (1944) *J. Phys. U.S.S.R.* **8,** 110 (paper T2 on p. 234).

LONDON, F. (1936) *Proc. Roy. Soc., London* A **153,** 576.

LONDON, F. (1938a) *Nature,* **141,** 644.

LONDON, F. (1938b) *Phys. Rev.* **54,** 947.

LONDON, F. (1939) *J. Phys. Chem.* **43,** 49.

LONDON, F. (1954) *Superfluids,* Vol. II, Wiley.

LONDON, H. (1946) *Report of Physical Society Cambridge Conference* **2,** 48.

LUCIAN OF SAMOSATA IIc. (A. D.) *A Truthful Tale,* book I.

MARTIN, P. and SCHWINGER, J. (1959) *Phys. Rev.* **115,** 1342.

MCMILLAN, W. L. (1965) *Phys. Rev.* **138** A 442.

MILLER, A., PINES, D. and NOZIÈRES, P. (1962) *Phys. Rev.* **127,** 1452.

ONSAGER, L. (1948) Unpublished remark at a Low Temperature Physics Conference at Shelter Island.

ONSAGER, L. and PENROSE, O. (1956) *Phys. Rev.* **104,** 576.

OSBORNE, D. V. (1950) *Prog. Phys. Soc.* A **63,** 909.

PESHKOV, V. P. (1944) *J. Phys. U.S.S.R.* **8,** 131.

PESHKOV, V. P. (1946) *J. Phys. U.S.S.R.* **10,** 389 (paper E12 on p. 166).

PESHKOV, V. P. (1960) *J.E.T.P.* **38,** 799. *Soviet Phys. J. E. T. P.* **11,** 580.

PETRU, Z. (1968) *Acta Phys Polon.* **35,** 49.

PITAEVSKI, L. (1956) *J.E.T.P.* **31** 536 (paper T7 on p. 335).

RAYFIELD, G. W. and REIF, F. (1963) *Phys. Rev. Lett.* **11,** 305.

RAYFIELD, G. W. and REIF, F. (1964) *Phys. Rev.* **136** A 1194.

REEKIE, J. and HUTCHINSON, T. S. (1953) *Phys. Rev.* **92,** 827; see also L. GOLDSTEIN and J. REEKIE (1955) *Phys. Rev.* **98,** 857.

ROLLIN, B. V. and SIMON, F. (1939) *Physica* **6,** 219.

STERN, H. (1966) *Phys. Rev.* **147,** 94.

THOMSON, W. (1871) *Nature* **4,** 262.

TISZA, L. (1938) *Nature* **141,** 913.

TISZA, L. (1940) *J. Phys. at radium* (8), **1,** 164, 350.

TISZA, L. (1947) *Phys. Rev.* **72,** 838.

WAGNER, H. (1966) *Zf. f. Phys.* **195,** 273.

WHITMORE, S. C. and ZIMMERMAN, W. (1965) *Phys. Rev. Lett.* **15,** 389.

VAN URK, KEESOM, W. H. and KAMERLINGH ONNES, H. (1926) *Leiden Comm.* **179a.**

VINEN, W. F. (1961a) *Progress in Low Temperature Physics* (Editor GORTER), North-Holland, Amsterdam.

VINEN, W. F. (1961b) *Proc. Roy. Soc.* A **260,** 218.

VINEN, W. F. (1966) *Quantum Fluids* (Editor BREWER), North-Holland, Amsterdam.

Yang, C. N. (1962) *Rev. Mod. Phys.* **34**, 694.
Yarnell, J. L., Arnold, G. P., Bendt, P. J. and Kerr, E. C. (1959) *Phys. Rev.* **113**, 1379.
Yntema, J. L. and Schneider, W. G. (1950) *J. Chem. Phys.* **18**, 641, 646.
Ziman, J. M. (1953) *Proc. Roy. Soc.* A **219**, 257.
Zubarev, D. N. (1960) *Sov. Phys.-Uspekhi* **3**, 320.

PART 2. ORIGINAL PAPERS

Experimental Papers

EI. FURTHER EXPERIMENTS WITH LIQUID HELIUM*

V. The Variation of Density of Liquid Helium Below the Boiling Point

H. KAMERLINGH ONNES and J. D. A. BOKS

V. THE LIQUID DENSITY

§ 1. *Introduction.* Previously† the startling result was obtained that liquid helium, notwithstanding its simple chemical character, presents a maximum of density. It was of great interest to make a closer study of this phenomenon, and to obtain accurate data on the course of the density on descending temperature, which is also very remarkable in other respects. As regards the obtaining of temperatures in the neighbourhood of 2·3°K, the question is of importance, because, when the liquid is cooled by evaporation, the cold layer continues to float on top; also for the explanation of the very singular phenomena of evaporation, formerly found‡ the knowledge of density in dependence on temperature, is very desirable.

§ 2. *Description of the apparatus with which the measurements of the liquid density have been carried out.* The dilatometer used for this purpose, is represented separately in Fig. 1. The volume of reservoir *D* is about 12 cm³. It can be closed at *A* with a tap, which

* *Leiden Comm.* **170b** (1924).
† *Comm.* N°. 119.
‡ *Comm.* N°. 159, § 12.

is found under the helium in the bath.* The stem of the dilatometer is narrowed close above the reservoir at *B*.† The diameter of the upper part is 0·195 cm², that at *B* is 0·049 cm². This will render it possible to determine a variation of the density of the liquid helium accurately, when the dilatometer is filled up to *B*. When the tap is

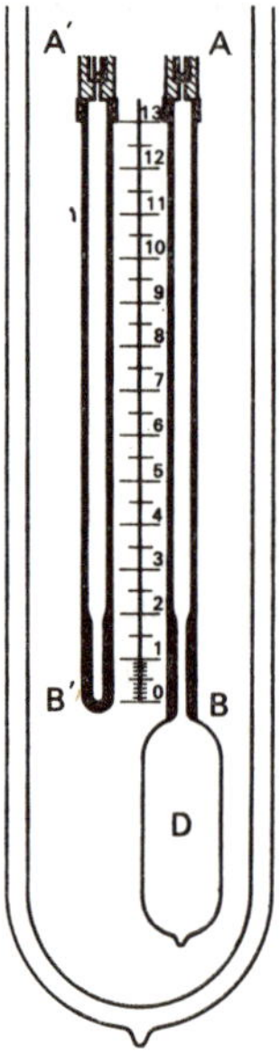

FIG. 1

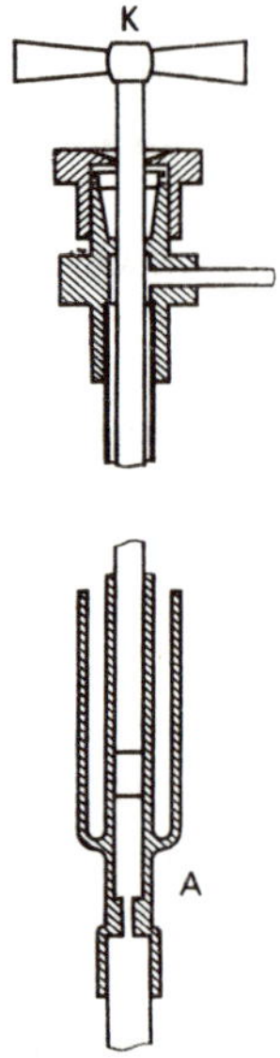

FIG. 2

closed at *A*, and the temperature of the bath is varied, the position of the meniscus of the helium in the dilatometer will change in consequence of the variation of the density of the liquid, but also of the variation of the density of the vapour above the liquid; in other words, through condensation or evaporation. This latter

* This tap can be regulated from the outside. The tap is separately represented in Fig. 2.

† Between line 1.5 and line 2.0 there have been erroneously drawn only three division lines. (Cf. Fig. 1).

might be approximately calculated, but not accurately, as the densities of saturated vapour are not accurately known. Therefore there has been applied a control tube beside the dilatometer, which agrees as closely as possible with the dilatometer, but which has no reservoir. It can be shut off at A' by a tap, and has a narrowing at B'. The diameter of the upper part is 0·199 cm^2, that at B' is 0·045 cm^2 (cf. Fig. 2). When now the meniscus of the helium is at B, and when there is also liquid in the control tube up to the same level, then the variation of the position of the meniscus in B can be read in B' for so far as it is the consequence of condensation or evaporation, through which a greater accuracy is effected. The variation of the density of the liquid in the control tube can be left out of consideration. In the bath is found the stirrer R, which can be worked by an electromagnet E moving up and down, by means of the piece of soft iron W (within the glass tube). The normal volume of the liquid helium was measured in the large iron cylinder F, the volume of which (about 12 litres) had been determined volumenometrically. This tube stood in water, which notwithstanding the water was stirred, gave rise to a great inaccuracy in the temperature, and consequently also in the normal volume, so that the absolute values of the liquid densities determined, partly also on account of the not very great accuracy with which the volume of the cylinder has been determined, are not very accurate. By the application of the taps at A and A' (see Fig. 1), by means of which the dilatometer and the control tube can be shut off, the great drawback which still existed in the apparatus like it was formerly used* is obviated, i.e. that along a large part of the capillary which contained the helium, the temperature was so uncertain.

§ 3. *The measurements.* These were carried out in the following way. At a temperature of about 4·2°K the dilatometer was filled to line 12, after which the tap K was closed. When the temperature

* *Comm*, N°. 119,

was lowered, the meniscus of the helium in the stem fell finally below the graduated part in *D*, and then appeared again. It was then necessary to add helium to keep the lowest position still in the narrow part *B*. Then the temperature was slowly lowered, in which the meniscus fell slowly, and remained in the part *B*. Now also the control tube in *B′* was filled with liquid, to about the same level as in *B*. By now lowering and raising the temperature more and more slowly, the lowest position could be measured accurately. The variation of the meniscus in *B′* now gave the change, which was only the consequence of condensation or evaporation. The control tube renders the accuracy of the *relative* values perfectly satisfactory; the *absolute* values of the numbers is, however, only accurate down to $\frac{1}{350}^{\text{th}}$ about, because the normal volume of helium, with which the measurement was made, could not be determined more accurately, as was already referred to in § 2. From the relative values a confirmation may be derived of the existence of a maximum of density in liquid helium, formerly found.* This maximum appears to lie at 2·29°K. Below this temperature the density seems to approach again to a constant value; an indication of this was also already found. The change of the density is about 1:140.

With the same filling were measured

1. the points 1, 2, 3 of May 5, 1922.
2. the points 4, 5, 6 and 7 of Febr. 2, 1922.
3. the points 1 to 10 inclusive of Febr. 19, 1922.
4. same filling 11, 12, 13 of Febr. 10, 1922.

The points of May 5 were calculated with a different normal volume. The points 4, 5, 6 and 7 of Febr. 2 were compared directly with the points of Febr. 10 by equating the density of the helium

* *Comm.* N°. 119, § 5.

at point 4 of Febr. 2 ($T = 2{\cdot}29$°K) to the density at point 1 of Febr. 10 ($T = 2{\cdot}29$°K), the points 11, 12, and 13 of Febr. 10 having been made to agree with the points 1 to 10 inclusive of the same day by taking the same temperature at point 11 as at point 9.

§ 4. *Results of the measurement of the liquid density.**

T	d†	ϱ‡	Date	N°
4·22 °K	698·7	0·1249	May 5, 1922	1
3.90	730·4	0·1305	May 5, 1922	2
3·30	776·6	0·1388	May 5, 1922	3
2·56	811·4	0·1450	Febr. 10, 1922	10
2·56	811·7	0·1451	Febr. 10, 1922	7
2·37	816·7	0·1459	Febr. 10, 1922	8
2·30	818·1	0·1462	Febr. 10, 1922	2
2·30	818·1	0·1462	Febr. 10, 1922	3
2·29	818·2	0·1462	Febr. 10, 1922	1
2·29	818·2	0·1462	Febr. 2, 1922	4
2·29	818·2	0·1462	Febr. 2, 1922	5
2·21	816·6	0·1459	Febr. 10, 1922	4
2·10	815·7	0·1458	Febr. 2, 1922	6
2·10	815·4	0·1457	Febr. 10, 1922	5
1·93	814·1	0.1455	Febr. 10, 1922	6
1·92	814·4	0·1455	Febr. 2, 1922	7
1·59	813·0	0·1453	Febr. 10, 1922	9
1·59	813·0	0·1453	Febr. 10, 1922	11
1·28	812·4	0·1452	Febr. 10, 1922	12
1·20	812·4	0·1452	Febr. 10, 1922	13

* Graphically represented in Fig. 3.

† The values in this column are equal to V_N/V_{liq}, in which V_N is the volume (under normal circumstances) of the helium, with which the measurement was made, and V_{liq} the volume occupied by the helium in the liquid state at the different temperatures.

‡ These values for ϱ_{liq} have been obtained by multiplying the values for d from the preceding column by 0·0001787, which value is given for $\varrho_{0°\text{C. } 760 \text{ mm}}$ in *Comm.* N°. 119.

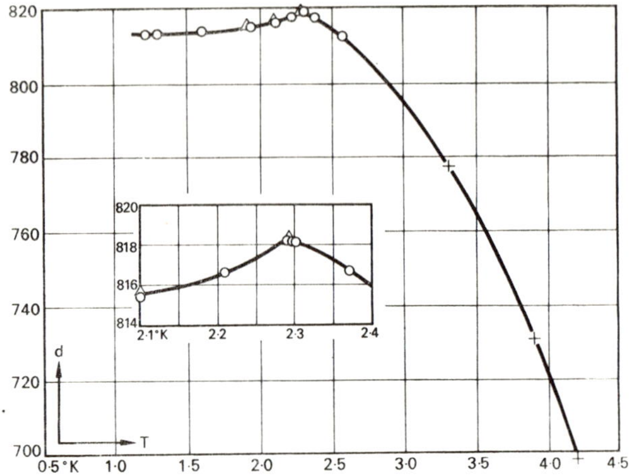

FIG. 3

In Fig. 3 the asymmetrical form of the curve is striking. It will be well to determine more points in the neighbourhood of the maximum.

E2. TWO DIFFERENT LIQUID STATES OF HELIUM*

W. H. KEESOM and M. WOLFKE

§ 1. When measuring the dielectric constant of liquid helium between the boiling point and 1·9°K on June 11th last, we observed† that at a temperature almost corresponding with that one at which KAMERLINGH ONNES had found a maximum in the density curve, the dielectric constant shows a sudden jump or at least a jump made in a very small temperature region. The thought suggested itself that at that temperature the liquid helium transforms into an other phase, liquid as well. If we call the liquid, stable at the higher temperatures "liquid helium I", the liquid, stable at the lower temperatures "liquid helium II", then the dielectric constant of liquid helium I should be greater than that of liquid helium II. Seeing that this result was only founded on one series of measurements we wished to repeat the measurements in order to get more security. Two attempts for that purpose failed (on June 17th and July 12th). At a repetition on July 19th the circumstances were less favourable, so that in our opinion the results, though they *did* point in the same direction as those of June 11th, did not yet give a sufficient affirmation. So we decided to wait for new measurements.

§ 2. Meanwhile our attention had been drawn by the following facts:

* *Leiden Comm.* **190b** (1927).
† *Comm.* N°. 190*a*.

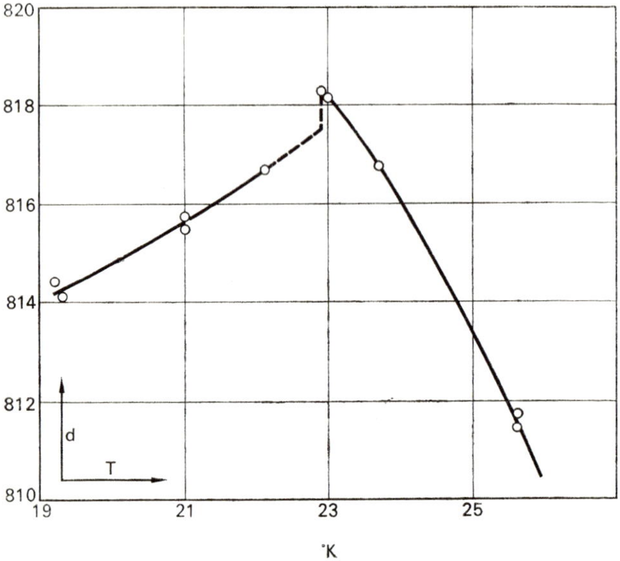

FIG. 1

a. The results of the density measurements of KAMERLINGH ONNES and BOKS* may be associated as well, if not better (see Fig. 1, which we must compare with Fig. 5 of Comm. N°. 170*b*) with the admission of a jump in the density at 38 mm helium pressure (to the amount of about 1‰, the density of liquid helium II smaller than that of liquid helium I) than with that of a smooth maximum at that place.

b. In the paper of DANA and KAMERLINGH ONNES† on the specific heat of liquid helium results are only mentioned at temperatures higher than the one expressed above. However, they have also experimented round that temperature, but then they found values, which did not agree with those measured at higher temperatures. For a part of the experiments this will be due to

* *Comm.* N°. 170*b*.
† *Comm.* N°. 179*d*.

condensation of the helium vapour, as DANA and KAMERLINGH ONNES point out. With some experiments leading to deviating results a suchlike condensation would not have been expected according to the data, which are kept in the archives of the laboratory.

In fact from three of those experiments we could deduce nearly corresponding values for the transformation heat with an average value:

liquid helium I = liquid helium II + 0·13 cal./gram.

c. The results of the measurements of DANA and KAMERLINGH ONNES* concerning the heat of vaporisation of liquid helium clearly point to a jump in the heat of vaporisation (see Fig. 3 there). They already say that it may be possible that those results should indicate "that near the maximum density something happens to the helium, which within a small temperature range takes place perhaps even discontinuously".† According to those results at the transformation point the heat of vaporisation of liquid helium II would be greater than that of liquid helium I and that to an amount which corresponds in order of magnitude with the value of the transformation heat given under *b*.

d. Measurements of VAN URK, KEESOM and KAMERLINGH ONNES‡ concerning the surface tension of helium seem to indicate clearly a jump in the value of the surface tension between 33 and 39 mm mercury pressure to the amount of about 3% (see Fig. 2 there) and in such a way that the surface tension of liquid helium II is smaller than that of liquid helium I.

§ 3. While the repetition, under improved circumstances, of the measurements of the dielectric constant mentioned in § 1, had to be put off on account of the absence of one of us from Leiden,§

* *Comm.* N°. 179*c*.

† *Comm.* N°. 179*c*, p. 31, note 1. The authors continue: "The change of density of the liquid also indicates something of the same kind".

‡ *Comm.* N°. 179*a*.

§ We hope that this repetition will soon take place.

the provisional value of the transformation heat mentioned in § 2*b* raised the surmise that the transformation of the liquid helium should show itself in the cooling or heating curve, when a quantity of liquid helium was being cooled regularly (by gradual lowering of the pressure) resp. heated itself in consequence of the supply of heat by conduction and radiation.

This experiment was carried out on November 18th last. The course of the pressure on a mercury manometer was thereby followed with the eye. With the cooling as well as with the heating a short halt of the manometer was stated each time at 39 mm pressure.

Besides a stirrer, with which the liquid was continually stirred during the experiment, the helium bath contained a constantin resistance. The intention was to register photographically by means of this the course of the temperature with the time. This, however, did not succeed on that day partly because the heating just like the cooling passed off sooner than we had expected, partly because the influence of a change in temperature of the leading wires to the constantin resistance (Diesselhorst galvanometer, Thomson bridge connection) did not appear to be sufficiently eliminated.

A repetition took place on Nov. 29th. Seven experiments were made: 4 coolings and 3 heatings.

The course of the mercury manometer showed again the same phenomenon as on Nov. 18th. Especially the course by heating was very characteristic. The manometer beginning from about 5 mm at first rose comparatively slowly, then stood still for a moment (always at the same point) and after that suddenly began to rise quickly. In one of these experiments, in which the manometer was constant for an estimated time of 4 sec., the pressure-reading was 38 mm.

During each of those experiments the deflection of the galvanometer in the Thomson bridge connection could be registrated.*

* We heartily thank Mr. J. N. van den Ende, phil. cand., for his assistance in these experiments.

On each of the thus obtained resistance-time-registrations, which can be almost considered as temperature-time-curves, the transformation point can be clearly recognized.

Visually nothing was to be seen of the forming of a second phase in the liquid helium. This need not astonish us as the refraction indices of the two phases will differ very little.*

From the fact that at lower temperatures the liquid is evidently being moved as easily by the stirrer as at higher it also appears that helium II is a very mobile liquid.

On December 7th measurements were made concerning the resistance of the constantin wire mentioned above, particularly near the transformation point, the pressure of the helium being kept constant during each measurement. We shall possibly revert to these experiments later. After that we had the opportunity to register still 3 cooling- resp. heating-curves in the way mentioned above. During these registrations the temperature changed considerably more slowly than at the preceding experiments. Though the then obtained curves show peculiarities, which we cannot yet quite interpret, still each of them shows very clearly that at a pressure of about 38 mm something very peculiar takes place.

On December 16th measurements were made by Dr. SOPHUS WEBER, Mr. NØRGAARD and one of us on the vapour pressure of helium particularly also near the point in question here. This time the temperature was measured with a helium thermometer of which the pressure was read by means of a hot wire manometer. The results of these measurements will be communicated later. Fig. 2 gives the successive readings of the galvanometer in the hot wire manometer connection, when after a series of these measurements the helium bath heated itself. The transformation point is very clearly indicated in this heating curve.

§ 4. Through the preceding experiments we think the fact to be established that at a pressure of about 38 mm a very peculiar

* Herewith should be compared what has been observed at the experiments concerning the solidification of helium: *Comm.* N°. 184*b*.

change appears in the liquid helium, which in any case takes place within a very small temperature region (0·08 of a degree). Though we have not been able to observe visually the two phases side by side, we yet think it most probable that we have to do here with

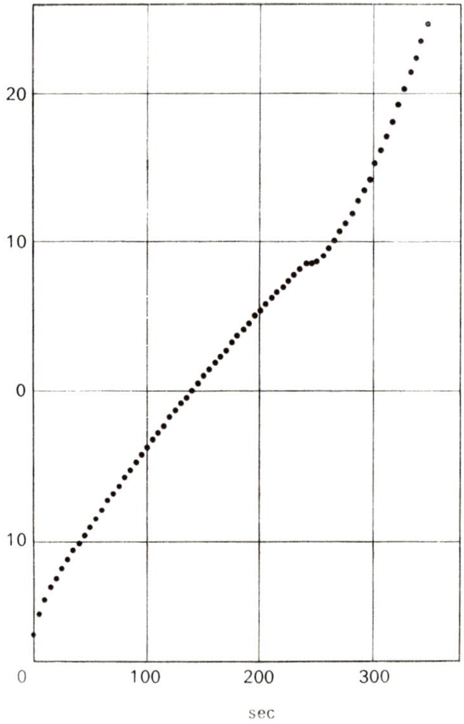

FIG. 2

two different states of liquid helium, which transform into each other at the temperature corresponding to the pressure mentioned. Of those two phases the liquid helium II (stable at the lower temperatures) compared with liquid helium I has:

a smaller density,
a greater heat of vaporization,
a smaller surface tension,

while the transformation liquid helium II → liquid helium I takes place with an absorption of heat, of which the amount can be valued for the present at 0·13 cal./gram.

We put the temperature, at which the transformation takes place, for the present at 2·3°K.*

It is remarkable that this transition takes place at a temperature, which, roughly speaking, may be considered to correspond with those at which other substances have their melting point.

So helium has a triplepoint liquid I-liquid II-vapour, which until now has only been found with a number of complex substances, which have a mesomorphic (liquid crystalline) phase. Whether the latter is also the case with helium, will have to appear from further experiments.

Likewise whether and, if so, in which point the transformation curve liquid I-liquid II meets the melting curve.

* It is evident that the deduction of the temperature from the value of the vapour pressure especially near this point, shall want a reconsideration.

E3. SPECIFIC HEAT OF LIQUID HELIUM*

W. H. KEESOM and K. CLUSIUS

Summary. We have measured the specific heat of liquid helium under its saturated vapour pressure between 1·4° and 4·1°K. The curve of the specific heat increases sharply starting at 1·4°K until it shows a steep maximum (molar heat about 9 cal) at 2·19°K; it then shows a shallow minimum at about 2·6°K, and after that in accordance with the results of Dana and Kamerlingh Onnes it increases slowly.

These measurements were complemented by a few observations of the specific heat at constant volume between 2·5° and 4·0°K.

1. *Introduction.* The specific heat of liquid helium has already been investigated. Dana and Kamerlingh Onnes† have given data about the specific heat under the saturated vapour pressure between 4·05° and 2·61°K and shown that in this temperature range the measurements can be represented in first approximation by a linear function.

It was desirable to extend our knowledge about this topic and to go to lower temperatures than it was possible for the above-mentioned authors. The circumstance that liquid helium apparently exists in two different states‡ which merge into one another at about 2·19°K particularly encouraged us.

* *Leiden Comm.* 219e.

† L. I. Dana and H. Kamerlingh Onnes, *Leiden Comm.* 179d.

‡ W. H. Keesom and M. Wolfke, *Leiden Comm.* 190b (reprinted in the present volume as paper E2 on p. 65.). We have changed the temperature of 2·30°K given in their paper using the temperature scale of W. H. Keesom, Sophus Weber, and G. Schmidt, *Leiden Comm.* 202c.

We could expect especially that by measuring the specific heat under higher pressure one could follow the shift in the transition point caused that way. With this aim in view we have constructed our calorimeter vessel in such a way that it could withstand the pressures we hope to use.

Due to circumstances we have not yet been able to perform measurements of the specific heat at higher pressures. On the other hand, the form of the transition curve has been established sufficiently clearly from the cooling curves taken at different pressures. The data obtained in this way have already been published.*

We had also hoped to extend the data we have obtained about the specific heat under saturated vapour by further measurements. As this planned extension of our work has been postponed several times we feel that we should no longer postpone publication.

2. *Description of the apparatus.* Figure 1 shows the main parts of our apparatus, the construction of which was helped by the experience in our institute about the measurement of specific heats at helium temperatures.† In a helium cryostat the calorimeter A is enclosed by a measurement cylinder B. A tube made of new silver N which widens at the top makes it possible either to evacuate the space between the measurement cylinder and the calorimeter or to fill it with some helium gas to provide heat exchange with the bath. The use of the metal cylinder gives the calorimeter the best possible protection against irradiation. To weaken also the energy incident through the new silver tube we had soldered at the widening of the tube a metal screen S which was also cooled to a low temperature by the bath. We used a three-stage Gaede pump, which was connected by a glass tube of 50 mm diameter to the upper end of the new silver tube, to evacuate the measurement space. The calorimeter itself consists of two parts: the core, which can be screwed in and which contains the elements for the input of

* W. H. Keesom and K. Clusius, *Leiden Comm.* 216b.

† W. H. Keesom and J. N. van den Ende, *Leiden Comm.* 203d.

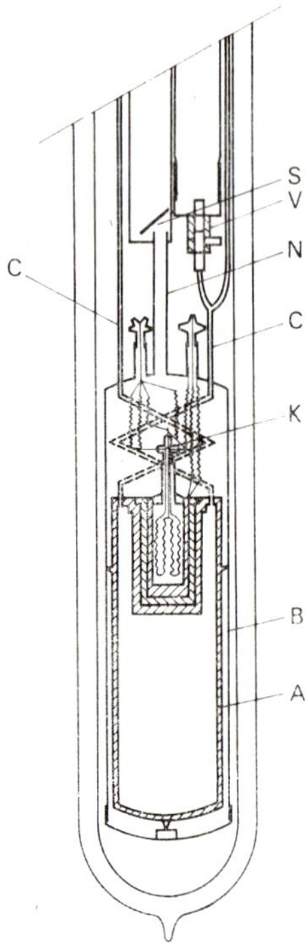

FIG. 1

the energy and to measure the temperature, and the calorimeter vessel proper which serves to take in the liquid helium.

The core was constructed in the same way as that part of the apparatus used by Keesom and van den Ende. However, it was improved in two ways: first of all we used for the temperature measurement for the first time a phosphor-bronze thermometer which is especially suited for the present case* and which can, if needed, be used with an altered constantan thermometer and which hung again in a helium atmosphere. Secondly, we used for the

* W. H. Keesom and J. N. van den Ende, *Leiden Comm.* 203c.

admission of the thermometer wire an elegant glass cross K which did not only save considerable space as compared with the arrangement used earlier but especially ensures a fast temperature equilibrium in this necessary appendix to the core.

The calorimeter consists of copper-lead of 1 mm thickness and has a weight of 180·9 g. It is fixed to a definite position in the measuring cylinder by means of three steel centring points in its upper third and a bottom point. The bottom point rests in a glass cage which is fixed to the bottom of the measuring cylinder. Two contracid capillaries C of about 0·8 mm outer diameter lead to the calorimeter; one of them is connected through a valve with the high-pressure tubing of the helium compressor while the other one is connected with a gate valve V which lies in the liquid helium.* Below this valve a branch tube leads to a precision manometer; the calorimeter itself was tested at a pressure of about 50 atm. The capillaries are led airtight through the new silver cover of the measuring vessel and soldered to the calorimeter after having been bent into a circular form to diminish the heat transfer between bath and calorimeter. In the cover we have soldered by means of nickel–steel bars which were melted in two glass crosses of a similar construction as the cross fixed to the core; they are used to pass through the heating, and potential and current wires.

To fill the calorimeter the helium bath was left to boil under diminished pressure. After all metal parts had been cooled to the appropriate lower temperature and by the opening of the gate valve the connection with the well-evacuated calorimeter had been established, the pressure over the helium surface was suddenly increased to 1 atmosphere. One could then observe that the helium meniscus began to fall rapidly and came to rest after about 30 sec: the liquid helium had then been pressed through the narrow capillaries and filled the calorimeter. To be absolutely sure that the

* Cf., for example, L. I. Dana and H. Kamerlingh Onnes, *Leiden Comm.* 179c.

filling was complete, one could then, after the gate valve is closed from the side of the compressor, give a slight overpressure (about 2 atm) to the helium and observe whether the manometer at the other capillary responded immediately. This was always the case. Then the pressure was released to reach the saturated vapour pressure and this was followed carefully while observing the calorimeter temperature and finally the valve was closed.

Evacuating provided very great difficulties. It turned out, however, that after a period of about 45 minutes of pumping during which the bath temperature all the time was kept as low as anywhere else (about 1·3°K), the temperature of the calorimeter became more and more clearly moving upwards. In the meantime the vacuum became steadily better (about 10^{-4} mm Hg); short heat pulses had as far as possible removed gas which possibly was absorbed on the calorimeter. The measurements could then begin and even at 1·5°K they gave useful and well-reproducible values.

3. *Calibration of the thermometer.* We carefully calibrated the thermometers in liquid and solid hydrogen using two platinum resistance thermometers and in liquid helium by carefully observing the vapour pressure; a Dieselhorst potentiometer provided excellent service for the resistance measurements. As the later measurements were made only at liquid helium temperatures and only using the sensitive phosphor-bronze thermometer we have restricted the data to those values.

To be sure, the temperatures were calibrated every day we performed measurements and we always found the same values. The temperature was derived using the recently determined vapour pressure curve of liquid helium.* The temperature dependence of the resistance of the thermometer used was—as had been found earlier—practically linear between 1·15° and 4·35°K. The considerable increase of measuring accuracy reached in the present paper

* W. H. Keesom, Sophus Weber, and G. Nørgaard, *Leiden Comm.* 202b; W. H. Keesom, Sophus Weber, and G. Schmidt, *ibid.* 202c.

TABLE 1. PHOSPHOR-BRONZE RESISTANCE 1930 II.

Day of measurement	Temp. °K	Resistance Ω	Day of measurement	Temp. °K	Resistance Ω
17. X. 1930	4·233	22·0240		2·917	20·3544
	3·591	21·1819		3·299	20·8124
	2·919	20·3606		3·586	21·1706
	2·516	19·9061		4·013	21·7292
	2·042	19·4072		4·351	22·1789
	1·524	18·9061		4·287	22·0959
	1·158	18·4953		4·219	21·9990
	1·473	18·8253	21. XI. 1930	4·220	22·0074
	1·831	19·2046		2·245	19·6160
	2·149	19·5001	27. XI. 1930	4·163	21·9330
	2·597	19·9909			

is mainly due to the use of this thermometer. The temperatures were read from a drawing which was drawn on a very large scale; all measuring points were lying beautifully on one curve; however, there seemed to be a slight difference between data taken at increasing and decreasing temperatures in as far as the first as compared to the latter corresponded to slightly larger values of the resistance, but nearly within the limits of the accuracy. This perturbation is, however, very small and in any case only leads to an adjustment of at most 0·010 degree to be made to the curve which we considered to be the correct one.

4. *Heating energy*. In the way indicated earlier the measurement was performed by observing the current I in the heating circuit, the potential drop E over the heating resistance using a precision ampere/voltmeter, and by a suitable choice of the heating time t which was determined by a clock with a second pendulum using a method which has been described earlier.* The heating resistance w was about 64·1 Ω at liquid helium temperature so that the heat

* W. H. Keesom and H. Kamerlingh Onnes, *Leiden Comm.* 143.

developed in the calorimeter in a measurement is given by

$$\{EI/[1+(w/w')]\}\,(t/4{\cdot}1824)\text{ cal.}$$

The term involving w' takes into account the current used by the potential measuring apparatus when its resistance together with the lead wires is w'. This correction is about 6% in the present case.

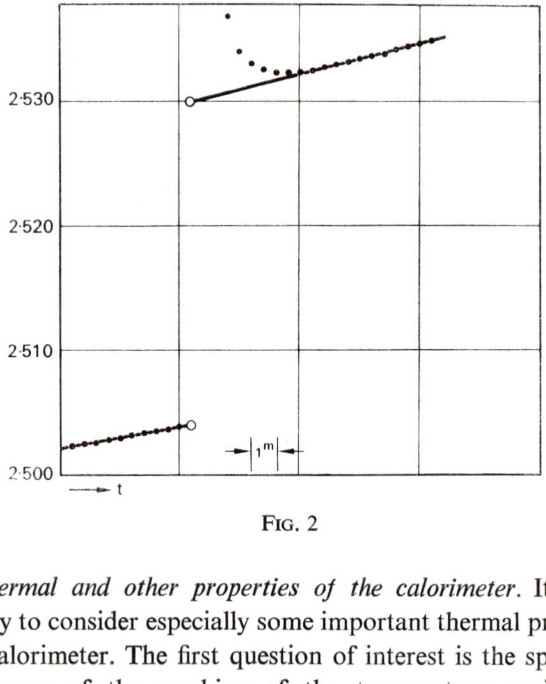

FIG. 2

5. *Thermal and other properties of the calorimeter*. It is now necessary to consider especially some important thermal properties of the calorimeter. The first question of interest is the speed and completeness of the reaching of the temperature equilibrium. From the description given here we had not prepared for the possibility to stir the contents of the calorimeter which can be done electromagnetically. Experience shows that the temperature had reached equilibrium after about 4 minutes below and at most 6 minutes above the density maximum. Figure 2 shows a typical, average good, observation of the way the temperature changes during a measurement. The abscissa gives the time, the ordinate

the reading at the compensation bank. One notices the linear increase in the initial period, the strong temperature effect of the main period with the following decrease in the temperature through the flowing away of the energy from the core to the calorimeter, and the final period. One might doubt whether the temperature equilibrium is reached sufficiently fast notwithstanding the excellent thermal conductivity of copper. However, in favour of this are the following facts: the calorimeter is heated at the top. At temperatures below the density maximum the heated liquid sinks down in the cooler part and encourages temperature equilibrium through mechanic streaming. Above the density maximum the lighter substance floats on the cooler part and the temperature equilibrium can only follow through conduction through the walls. If the reaching of the equilibrium were very poor the thermometer would indicate a too high temperature after the heating so that we would get a too low specific heat. However, the results of this paper agree well with those of Dana and Kamerlingh Onnes within the limits of accuracy, and in their experiments they had a built-in strirring so that partial overheating could not take place. One is thus led to the assumption that the reaching of temperature equilibrium which has been observed in our experiments was indeed sufficiently fast and *a fortiori* was sufficiently fast at temperatures below the density maximum.

Notwithstanding repeated attempts it turned out to be impossible to obtain the heat capacity of the empty calorimeter through measurements. A simple calculation shows indeed that the calorimeter and the core which was also constructed mainly from copper and which weighed 77·9 g together represent a heat content of $3{\cdot}5\times10^{-3}$ cal at about 4°K, provided the Debye T^3 law retains its validity. At about 2°K only about 1600 kg copper would have the same heat capacity as the vessel filled with helium. We are thus dealing with an ideal calorimeter the heat capacity of which can practically be put equal to zero. One possible error must, however, be discussed briefly: a certain heat effect will be shown by the core which is filled at room temperature with helium at 100 cm pressure* if the helium were liquid at a certain temperature and evaporated during the next heating period. One can, however, estimate that this effect would not occur until 1·5°K and about 5 mm pressure.†

To measure the amount of helium we need to know apart from the temperature at which the filling takes place and the density determined by it the exact volume of the calorimeter. This was determined at room temperature through an exact volumenometric measurement using a comparison vessel of about the same content which was weighed filled with mercury, and using the linear expansion coefficient of copper‡ to find the helium temperature value. From three independent sets of measurements at 18°C the value of the volume was found to be 81·83 cm^3 after all necessary corrections had been made; the probable error was about $\pm$ 0·10 cm^3. By multiplying with the factor 0·99045 we obtained the volume at 4°K to be 81·05$\pm$0·10 cm^3.

It follows from this kind of measurement of the amount of helium that we did not take into account the liquid which remained in the capillaries. A rough estimate shows that the resultant error is 0·05% in the least favourable case so that we can completely neglect it in comparison with the random errors.

6. *Heat losses and other corrections.* It followed from the linear dependence of the temperature data taken every 30 seconds that one can assume that the heat transfer to the surroundings are regular and unaffected by the heating period so that we can apply the corrections unambiguously.

* It would have been better to have had a lower pressure during filling; this would not have affected the heat transfer.

† Later measurements of the heat capacity of the heating core (see W. H. Keesom and J. N. van den Ende, *Leiden Comm.* 219b, fig. 8) have shown that this effect is only about 1/500 of the heat capacity of the liquid helium in the calorimeter vessel.

‡ W. H. Keesom, F. P. G. A. J. van Agt, and Miss A. F. J. Jansen, *Leiden Comm.* 182a.

It is true that the values obtained in this way do not always give us directly the specific heats under the saturated vapour since the calorimeter is usually filled only partly due to thermal contraction. If one excludes the possibility of the formation of the vapour phase by using overpressure, one obtains the specific heat c_v and not data from which one can derive c_s in a simple manner. We have therefore used the vapour pressure measurements of Keesom, Weber, Nørgaard, and Schmidt to correct the vapour volume, as was done by Dana and Kamerlingh Onnes. However, this correction was only important for the second set of measurements, since the first set had been obtained through filling in the neighbourhood of the density maximum so that no appreciable vapour volume appeared.

7. *Experimental procedures.* After we had cooled down the calorimeter successively with liquid air, liquid hydrogen, and liquid helium, it was filled under exact observation of the temperature in the way already described and was cooled down as far as possible. The lowest temperature which we could reach was not below 1·3°K because of the many metallic leads going into the cryostat.

In each cooling process we measured a cooling curve to get a purely qualitative idea about the behaviour of the specific heat. After three-quarters of an hour of pumping we had attained a sufficient vacuum so that we could proceed with the successive heatings to measure the specific heat.

8. *Results and discussion.* (a) In Fig. 1 of *Leiden Comm.* 216b one of the cooling curves obtained by us has already been given. It can be concluded from the shape of the curve that below 2·19°K the specific heat must be larger than above it. As is shown by a comparison with other results obtained, the cooling curve indeed reflects the behaviour of the specific heat in all details.

The results of two measuring days are combined in Fig. 3 for the specific heat c_s and all essential data are given in Tables 2 and 3. The first thing to be noticed is the sharp maximum at 2·19°K. While all anomalies observed so far in liquid helium occur as relatively slight effects, the measurement of the specific heat shows itself just especially suited to provide an indication of the energetic conditions in this region.

One can see that the accuracy is excellent down to the lowest temperatures which must be ascribed primarily to the use of the phosphor-bronze thermometer as its excellent, uniform sensitivity allows us to measure even temperature increases of 0·05 degree

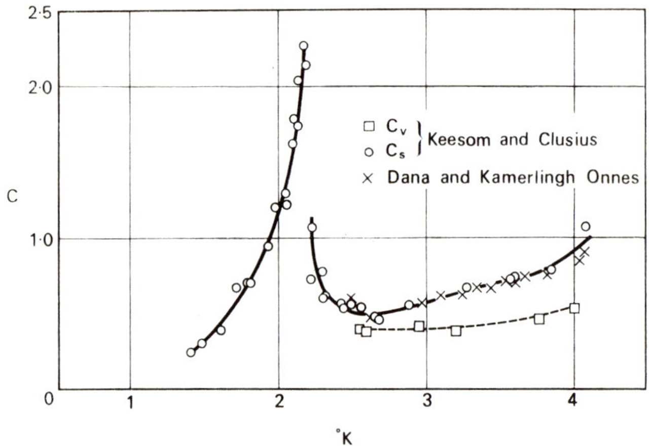

FIG. 3

with an accuracy of about 1%. This fact was particularly favourable for the determination of the data near the maximum. We consider the measurements to have an accuracy of about 3%; however, the results below 1·6°K may have errors which are twice as large.

A steep increase to molar heats of about 9·5 cal at 2·19°K is followed by a sudden drop to about one-third of this value, which is followed after a clear minimum at 2·65°K by a practically linear increase. The measurements between the maximum and the minimum show clearly a larger scatter than one might expect. One gets the impression as if the values found there depend strongly on

TABLE 2. CORRECTED ENERGY INPUT

4 December 1930 Number of experiment	Heating current I in mamp	Heating time t in sec	Heating potential in volt	Corrected energy input in watt	Average temperature $\bar{T}$ °K	Temperature difference ΔT °K	Heat capacity of the system cal/°K	Specific heat c_s in saturated state cal/°K, g
1	48·3	30	3038	0·1423	1979	0·072	14,160	1·212
2	48·3	8	3046	0·1426	2223	0·032	8 517	0·729
3	48·0	18	3014	0·1403	2298	0·085	7 096	0·608
4	48·2	24	3022	0·1413	2420	0·122	6 637	0·568
5	48·5	18	2995	0·1418	1720	0·077	7 915	0·678
6	48·4	24	3026	0·1421	2058	0·057	14,282	1·223
7	47·8	12	2987	0·1385	2114	0·019	20,883	1·788
8	48·0	20	3003	0·1397	2144	0·028	23,841	2·041
9	47·8	20	2995	0·1389	2176	0·025	26,522	2·271
10	48·3	30	3026	0·1418	2235	0·081	12,537	1·073
11	47·8	36	2987	0·1385	2440	0·187	6 365	0·544
12	48·0	36	3022	0·1407	2682	0·219	5 523	0·473

TABLE 3.

11 December 1930. Number of experiment	Heating current I in mamp	Heating time t in sec	Heating potential in volt	Corrected energy input in watt	Average temperature $\bar{T}$ °K	Temperature difference ΔT °K	Heat capacity of the system cal/°K	Correction for vapour volume per g	Specific heat c_s in saturated state cal/°K, g
1	43·6	12	2685	0·1101	1404	0·123	2 565		0·252
2	43·6	16	2680	0·1098	1478	0·133	3 142		0·308
3	43·0	20	1668	0·1078	1609	0·128	4 025		0·395
4	43·0	24	2666	0·1077	1790	0·086	7 184		0·705
5	43·2	32	2678	0·1087	1930	0·086	9 667		0·949
6	43·2	46	2680	0·1088	2047	0·090	13,288		1·304
7	42·8	34	2646	0·1064	2136	0·049	17,646		1·732
8	43·1	42	2680	0·1084	2192	0·050	21,788		2·139
9	43·0	42	2666	0·1077	2300	0·066	8 191	−0·015	0·788
10	43·2	32	2682	0·1089	2487	0·140	5 947	−0·019	0·565
11	43·3	72	2682	0·1096	1808	0·258	7 277		0·714
12	43·3	170	2684	0·1092	2098	0·273	16,492		1·619
13	43·3	42	2690	0·1095	2334	0·164	6 699	−0·015	0·642
14	43·25	60	2692	0·1094	2564	0·272	5 767	−0·019	0·546
15	43·25	76	2694	0·1095	2888	0·343	5 798	−0·011	0·558
16	43·3	78	2698	0·1098	3280	0·296	6 913	+0·003	0·682
17	43·25	78	2700	0·1098	3588	0·276	7 411	+0·016	0·743
18	43·2	66	2692	0·1094	3865	0·225	7 659	+0·034	0·786
19	43·1	66	2684	0·1087	4098	0·167	10,266	+0·069	1·076

accidental circumstances. This uncertainty disappears again above 2·6°K.

The behaviour of the specific heat below 2·19°K is very noticeable. One can represent the specific heat of the saturated liquid between 1·4 and 1·8°K approximately by 0·066 T^4. Above 1·8°K it increases more steeply. It would be of great interest to measure in this region the specific heat at constant volume.

We have also indicated in the figure the data of Dana and Kamerlingh Onnes, recalculated using the new temperature scale. These values, obtained in a completely different way, agree well with ours. It is interesting that, for instance, the increase at the lowest temperature reached by these authors is in fact appearing.

(b) Especially interesting is the question whether or not one can ascribe the effects occurring at 2·19°K to a simple phase transition. To determine this we added to the calorimeter put in a vacuum very small amounts of energy while watching the temperature.* The result was unambiguous: each energy input was always followed by a temperature increase. This increase became very small near 2·19°K and above it became again larger, but it was *always finite*. From this it follows that we do *not* have here a real transition point with an appreciable latent heat in the sense of Gibbs' phase rule.

We have tried to corroborate in a different way our conclusion that there can not be an appreciable latent heat at 2·19°K. If one measures the specific heat over a large interval including the maximum, the value obtained in that way must agree with the result of integrating graphically the detailed results obtained in the same temperature interval, provided that there is no appreciable latent heat. The experiments showed that between 1·961° and 2·234°K

* Compare the measurements 7, 8, and 9 of 4th December 1930 (Table 2) and 6, 7, 8, and 9 of 11th December 1930 (Table 3). Moreover, we went in a preliminary experiment through the specific heat maximum using heat pulses which were about 10 times smaller and we observed always a temperature increase.

the heat content was 0·442 cal/g, while graphical integration led to 0·441 cal/g, as was to be expected from the situation as we have described it.

Although a jumplike transition connected with a thermal effect does not appear to occur, the extremely sharp maximum in the specific heat curve indicates that this phenomenon occurs mainly in a very narrow temperature interval. This corresponds to the fact that several properties of liquid helium show a change* which is very similar to a jump. In that sense we could still consider the observed effect to be a transition from a state I to a state II.

(c) Even though the nature of the transition has become a little clearer in the above discussion, one must admit that its inner causes are still completely hidden from us. It does not seem very likely that a simple association effect or something like that can be the cause. Similar phenomena as shown here by liquid helium have been observed for certain solids† such as the ammonia salts,‡ solid hydrogenhalogenides§, and methane.|| One finds then internal transitions which are not sharply fixed to one temperature, but are spread over a somewhat larger temperature range. Pauling†† has suggested as an explanation of these phenomena in some cases that when the temperature is lowered the molecules, or at least parts of the molecules, which originally could rotate freely in the lattice slowly lose their rotational degrees of freedom in favour of the oscillational frequencies. Even though the assumption of rotations has no meaning for helium, there seems to be a close relationship between the effects observed in helium and the anomalies occurring in solids in as far as the states occurring in the

* Compare W. H. Keesom and M. Wolfke, *Leiden Comm.* 190b; paper E2 in this volume.

† A similar behaviour is also shown by ferromagnetics at the Curie point.

‡ F. Simon, Cl. von Simson, and M. Ruhemann, *Z. Phys. Chem.* **129**, 339 (1927).

§ A. Eucken and E. Karwat, *Z. Phys. Chem.* **112**, 467 (1924).

|| K. Clusius, *Z. Phys. Chem.* **B3**, 41 (1929).

†† L. Pauling, *Phys. Rev.* **36**, 430 (1930).

TABLE 4.

Date	Number of experiment	Helium content in calorimeter in g	Concentration mole/litre	Heating current I in mamp	Heating time t in sec	Heating potential in volt	Average temperature $\bar{T}$ °K	Temperature difference ΔT K°	Heat capacity of the system cal/°K	Specific heat c_v cal/°K, g
4·12·30	1	11·679	36·02	48·3	36	3030	2960	0·248	4·92	0·421
18·12·30	2	12·98	40·0	47·3	30	2924	2553	0·179	5·21	0·401
	3	12·98	40·0	47·0	50	2920	2956	0·291	5·30	0·408
	4	13·12	40·5	45·9	36	2904	2588	0·214	5·15	0·392
	5	13·12	40·5	47·4	47	2928	3213	0·291	5·15	0·392
	6	13·12	40·5	46·3	56	2878	3781	0·271	6·17	0·470
	7	11·92	36·8	46·7	36	2898	4007	0·167	6·55	0·549

Note: During experiment 1 the resistance in the potentiometer circuit was 2000 ohms, during the other experiments it was 1000 ohms.

liquid above and below 2·19°K show an energetic difference. At any rate the effect observed here shows clearly the necessity to perform measurements at extremely low temperatures also for those substances which in principle have a very simple structure.

(d) We can complement our results by a few data obtained for the specific heat c_v of liquid helium above 2·5°K. These measurements have also been incorporated in Fig. 3 and Table 4, but due to a certain uncertainty in the determination of the amount of helium they are not so reliable as the measurements of c_s. The amount of helium was determined by determining the compressibility of helium at 2·71°K* and by estimating the extra amount of helium from the filling pressure. The values of c_v, the pressure dependence of which could not be taken into account, increase slightly with temperature and thus show a behaviour completely similar to that of the values of c_v in hydrogen in the region where it is liquid.† Helium seems to be the only liquid except hydrogen for which such a quantum effect can be demonstrated. At least one could not observe an increase in the values of c_v just above the melting point for neon.‡

* See W. H. Keesom and K. Clusius, *Leiden Comm.* 219f.

† A. Eucken, *Verh. Dtsch. Phys. Ges.* **18**, 4 (1916); W. H. Keesom and H. Kamerlingh Onnes, *Leiden Comm.* 153a.

‡ K. Clusius, *Z. Phys. Chem.* **B4**, 1 (1929).

E4. ON THE HEAT CONDUCTIVITY OF LIQUID HELIUM

W. H. KEESOM and Miss A. P. KEESOM*

IN Commun. Kamerlingh Onnes Lab., Leiden, No. 235*d* § 6[1] we concluded that there is a sudden change in the heat conductivity when passing from He II to He I. We presumed that this change in the heat conductivity is rather important and we announced that measurements on this point were being prepared.

It now appears that B. V. Rollin[2] had come to the same conclusion and planned to determine the heat conductivity of liquid helium too.

Although the calculations of the experiments we made and even our programme of experiments to be made are not yet finished, the interesting results we obtained may justify a short preliminary report.

In our first experiments we examined the heat conduction in a layer of liquid helium which had the form of a circular cylinder, radius 18, height 5 mm. The liquid helium layer was contained between two copper pieces each containing a heating unit and a phosphorbronze thermometer. We only mention the result of the following measurement:

$$T = 3.3°\text{K}, \quad \varkappa = 6 \times 10^{-5} \text{ cal/degr. cm sec.}$$

This value is of the order of magnitude of the heat conductivity of gases at ordinary temperatures.

At temperatures below the lambda-point the heat conductivity

* *Physica* **3**, 359 (1936).

appeared to be too large to be measured in this way. When heating one copper piece the temperature of the other followed immediately without a measurable temperature difference. Further it appeared that after heating was stopped the temperature of both copper pieces most rapidly fell off again to that of the surrounding liquid helium bath. Evidently the heat flowed off through the capillary which served to fill in the liquid helium. As the material of the capillary was a bad heat conductor we concluded that the heat conduction took place through the liquid helium "wire" filling up the capillary.

For the next experiments we designed two copper pieces each containing a phosphorbronze thermometer and a liquid helium chamber of about 2 cm^3. The two helium chambers were connected by a capillary of non-conducting material. One of the copper pieces was provided with a heating unit. From the helium chamber of the other a capillary started through which the whole could be filled with liquid helium. Of course the copper pieces and connecting capillary were arranged in a space which could be evacuated. In these experiments it appeared that still the temperature of the two copper pieces followed each other too rapidly to make good measurements possible. To approve the situation we then connected the copper piece which wore the filling capillary by means of two strong copper strips to the surrounding mantle which is directly immersed in the liquid helium bath.

To eliminate the thermal resistance occurring at the limiting surface between copper and liquid helium II, and which appeared to be relatively very considerable, we made experiments with two different connecting capillaries, one long 22, the other long 94 cm. Internal diameter in each case 0·6 mm.

The measurements till now show some particularities which must be examined further. We restrict ourselves in this preliminary report to the statement that we measured at 1·4 and at 1·75°K values of the heat conductivity of liquid helium II of about 190 cal/degr. sec cm.

We note that this heat conductivity is about 200 times that of copper at ordinary temperatures, or about 14 times that of very pure copper at liquid hydrogen temperatures. Hence liquid helium II is by far the best heat conducting substance we know.

It appears further that the heat conductivity of liquid helium II at the temperatures mentioned is about 3×10^6 times that of liquid helium I.

Connecting this with the abrupt change of the heat conductivity in passing the lambda-point we may perhaps be justified in calling liquid helium II *supra-heat-conducting*.

REFERENCES

1. *Physica,'s-Grav.* **2**, 557, (1935).
2. Cf. *Physica,'s-Grav.* **3**, 269, note** (1936).

E5. HEAT CONDUCTION IN LIQUID HELIUM*

J. F. ALLEN, R. PEIERLS and M. ZAKI UDDIN

IT WAS found by Rollin[(1)] and by Keesom and Keesom[(2)] that liquid helium II shows an extraordinarily high heat conductivity, of the order of 190 cal/deg. cm. sec. We have measured the heat conductivity with the following method, which is essentially a combination of a vapour-pressure thermometer with a liquid-helium manometer. (The method was developed in discussion with Dr. J. D. Cockroft.)

An evacuated glass vessel is immersed in a helium bath. It contains a vertical cylindrical bulb, the bottom of which communicates with the outer bath through a glass capillary. The bulb also contains a heater wire through which a current can be passed. This current is so adjusted that part of the helium in the bulb evaporates, and only part of the bulb is filled with liquid. The level of the liquid is observed in a telescope and its height compared with the level of the bath. In equilibrium, the difference in height is such that the difference in hydrostatic pressure is equal to the difference in vapour pressure. In this way one obtains the temperature difference between the bulb and the bath, and this is practically also the temperature difference between the two ends of the capillary. For example, at 2·10°K, a level difference of 1 cm corresponds to a temperature difference of $1 \cdot 32 \times 10^{-3}$ deg. Since, on the other hand, the heat flow through the capillary is equal to the heat produced in the wire, one can determine the conductivity.

* *Nature* **140**, 62 (1937).

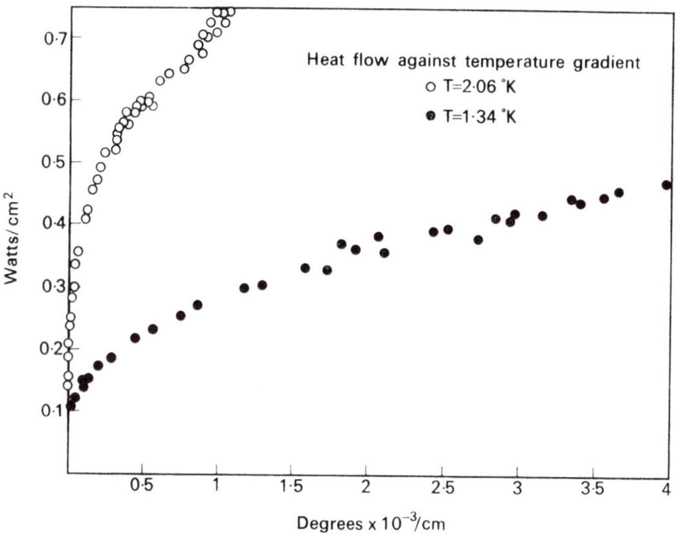

FIG. 1

It is essential for this method that one should actually obtain equilibrium, but it is easily estimated from the observed conductivity that equilibrium should be established in a fraction of a minute, and it was indeed observed that the level followed a change in the current very rapidly and reversibly. It is also easy to see that the heat conduction in the glass, in the leads carrying the current, etc., is negligible. In order to avoid convection, the warm end of the capillary was arranged to be lower than the cold end (the expansion coefficient of helium is negative).

Preliminary experiments showed the unexpected result that the conductivity depends on the temperature gradient. We therefore used an arrangement in which the level difference could vary from 12 cm. to about 0·3 mm, and measured the heat flow as a function of the temperature difference ΔT between the ends of the capillary. In this case the capillary was 16·8 cm long and had an internal diameter of 1·45 mm. The results are shown in Fig. 1, for bath temperatures of 2·06° and 1·34°K. It is clear that the heat flow is far from being proportional to ΔT, even for the lowest temperature differences we used. The ratio of heat flow to temperature gradient, which one may call conductivity, increases with decreasing ΔT. It reaches the value $1{\cdot}1 \times 10^4$ cal/deg cm sec at $\Delta T = 5 \times 10^{-5}$ and $T = 2{\cdot}06$. It appears, however, to rise still further and a refinement of the method is required to decide whether it tends to a finite value as ΔT tends to 0.

Qualitative measurements at a number of different temperatures between 1·34° and 2·19° showed the same general behaviour. At constant ΔT, the conductivity is relatively small near the λ-point (2·19°), reaches a maximum in the neighbourhood of 2·0° and decreases again at lower temperatures.

We cannot suggest an explanation for these results, for which there is no analogy in the behaviour of other substances. It seems impossible, however, to account for them by any errors in the experimental method; the effect seems actually to be a property of the liquid helium II.

REFERENCES

1. *Physica*, **2**, 557 (1935).
2. *Physica*, **3**, 359 (1936).

E6. NEW PHENOMENA CONNECTED WITH HEAT FLOW IN HELIUM II*

J. F. ALLEN and H. JONES

SINCE publishing the preliminary measurements on thermal conduction in liquid helium II[1] which showed a variation of conductivity with temperature gradient, further data have been collected concerning the phenomenon.

It was noted that the conductivity appeared to increase as the bore of the conduction capillary decreased. It was then found that at very low temperatures (1·08°K) small heat flows produced a rise in the level of the liquid in the closed bulb at the heated end of the capillary, instead of the fall caused by increased vapour pressure which was observed for higher heat flows. The apparatus was modified (Fig. 1) so that the top of the bulb at the hot end of the capillary was open to the vapour above the surface of the liquid helium bath. There was thus no possibility of a difference in vapour pressure between the liquid at either end of the capillary. As in the previous cases, heat was introduced electrically by means of a coil inside the bulb. On heating, the level of the liquid again rose in the bulb. The rise in level increased with increasing heat flow, and for constant heat flow increased with decreasing temperature. This effect opposes the effect of increased vapour pressure in the closed bulb, and thus will probably explain the apparent non-linear variation of heat flow with temperature gradient. More complete data on this phenomenon will be published later by one of us (J. F. A.).

* *Nature* **141,** 243 (1938).

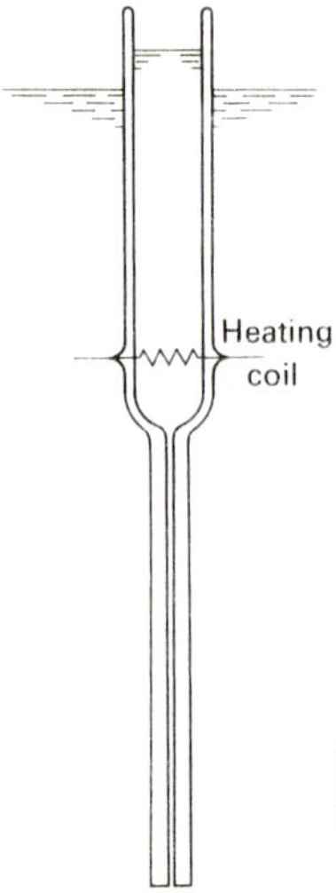

FIG. 1

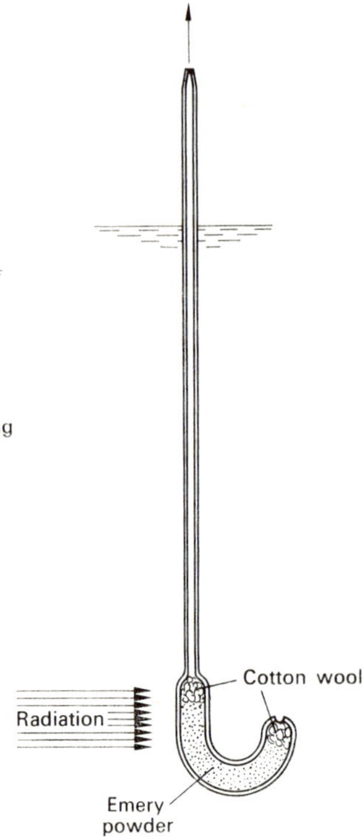

FIG. 2

A more striking manifestation of the above effect was observed by one of us (J. F. A.) in collaboration with A. D. Misener. Observations were being made on the flow of liquid helium II through a tube packed with fine emery powder (Fig. 2). The top of the tube was allowed to project several centimetres above the level of the liquid helium bath, and an electric pocket torch was flashed on the lower part of the tube containing the powder. A steady stream of liquid helium was observed to flow out of the top of the tube as long as the powder was irradiated.

The principle is the same as that for the thermal conductivity apparatus, in which heat is introduced electrically and flows through a capillary. Here the capillary is replaced by the minute interstices in the powder, and the heat input is obtained by means of the absorption of radiation by the powder particles. The apparatus forms, in effect, a very simple and efficient helium pump. The height of the jet becomes greater for lower bath temperatures and for powder particles of smaller size. The stream rises approximately 4 cm. above the jet, the top of the stream striking the side wall of the flask. The illumination for the photograph consisted of a 60-watt bulb, which also provided the radiant thermal energy for operating the helium pump. The maximum height of jet produced was 16 cm.

The significance of these results may perhaps be understood in the following manner. The observed specific heat of helium II shows that some cooperative phenomenon is concerned. According to Fröhlich[(2)], this is a disordering process analogous to that occurring in β-brass. Helium II is exceptional, however, in that almost the whole of the thermal energy at the λ-point is energy of disorder. We are led, therefore, to consider heat conduction as a flow of disorder rather than as a flow of vibrational energy. According to a suggestion of London's[(3)], the ordered state forms a diamond lattice which may be regarded as a body-centred cubic lattice in which only half the lattice points are occupied. According to Fröhlich, the disorder arises when some atoms fill the unoccupied sites of this underlying body-centred structure. Such "wrong" atoms would be able to jump from one possible site to the next. This model, admittedly naïve, contains an interesting feature, namely, that as the flow of disorder is weakened by collisions with the walls of the containing tube, momentum in the direction of flow is lost. This should be contrasted with conduction by vibrational waves which do not carry momentum. The reaction due to loss of momentum gives rise to a pressure in the opposite direction to the heat flow.

This process of conduction by flow of disorder has been considered in rather more detail by one of us (H. J., to appear elsewhere), and a second process by which a flow of disorder can be dissipated, namely, by interactions with vibrational waves, has been discussed. This latter process will not, of course, result in a "back pressure". This pressure will therefore be greater the greater the part played by the walls in the heat resistance. Lessening the bore of the tube in which heat flows, or filling with powder, increases, therefore, for a given heat flow, the magnitude of the back pressure.

REFERENCES

1. ALLEN, PEIERLS, and UDDIN, *Nature*, **140,** 62 (1937).
2. FRÖHLICH, H., *Physica*, **4,** 639 (1937).
3. LONDON, F., *Proc. Roy. Soc.*, A, **153,** 576 (1936).

E7. VISCOSITY OF LIQUID HELIUM BELOW THE λ-POINT*

P. KAPITZA

THE abnormally high heat conductivity of helium II below the λ-point, as first observed by Keesom, suggested to me the possibility of an explanation in terms of convection currents. This explanation would require helium II to have an abnormally low viscosity; at present, the only viscosity measurements on liquid helium have been made in Toronto[1], and showed that there is a drop in viscosity below the λ-point by a factor of 3 compared with liquid helium at normal pressure, and by a factor of 8 compared with the value just above the λ-point. In these experiments, however, no check was made to ensure that the motion was laminar, and not turbulent.

The important fact that liquid helium has a specific density ϱ of about 0·15, not very different from that of an ordinary fluid, while its viscosity μ is very small comparable to that of a gas, makes its kinematic viscosity $\nu = \mu/\varrho$ extraordinary small. Consequently when the liquid is in motion in an ordinary viscosimeter, the Reynolds number may become very high, while in order to keep the motion laminar, especially in the method used in Toronto, namely, the damping of an oscillating cylinder, the Reynolds number must be kept very low. This requirement was not fulfilled in the Toronto experiments, and the deduced value of viscosity thus refers to turbulent motion, and consequently may be higher by any amount than the real value.

* *Nature* **141**, 74 (1938).

The very small kinematic viscosity of liquid helium II thus makes it difficult to measure the viscosity. In an attempt to get laminar motion the following method (shown diagramatically in the accompanying illustration) was devised. The viscosity was measured by the pressure drop when the liquid flows through the gap between the disks 1 and 2; these disks were of glass and were optically flat, the gap between them being adjustable by mica distance pieces. The upper disk, 1, was 3 cm in diameter with a central hole of 1·5 cm diameter, over which a glass tube (3) was fixed. Lowering and raising this plunger in the liquid helium by means of the thread (4), the level of the liquid column in the tube 3 could be set above or below the level (5) of the liquid in the surrounding Dewar flask. The amount of flow and the pressure were deduced from the difference of the two levels, which was measured by cathetometer.

The results of the measurements were rather striking. When there were no distance pieces between the disks, and the plates 1 and 2 were brought into contact (by observation of optical fringes, their separation was estimated to be about half a micron), the flow of liquid above the λ-point could be only just detected over several minutes, while below the λ-point the liquid helium flowed quite easily, and the level in the tube 3 settled down in a few seconds. From the measurements we can conclude that the viscosity of helium II is at least 1,500 times smaller than that of helium I at normal pressure.

The experiments also showed that in the case of helium II, the pressure drop across the gap was proportional to the square of the velocity of flow which means that the flow must have been turbulent.* If, however, we calculate the viscosity, assuming the flow to have been laminar, we obtain a value of the order 10^{-9} C.G.S., which is evidently still only an upper limit to the true value. Using this estimate, the Reynolds number, even with such a small gap,

* In laminar flow the average velocity is proportional to the pressure difference and in turbulent flow it is proportional to the square root of the pressure difference *(Note by Editor.)*

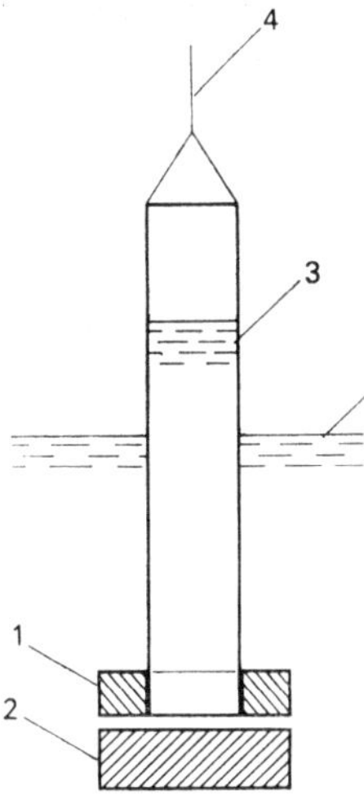

FIG. 1

comes out higher than 50,000 a value for which turbulence might indeed be expected.

We are making experiments in the hope of still further reducing the upper limit to the viscosity of liquid helium II, but the present upper limit (namely, 10^{-9} C.G.S.) is already very striking, since it is more than 10^4 times smaller than that of hydrogen gas (previously thought to be the fluid of least viscosity). The present limit is perhaps sufficient to suggest, by analogy with supraconductors, that the helium below the λ-point enters a special state which might be called a "superfluid".

As we have already mentioned, an abnormally low viscosity such as indicated by our experiments might indeed provide an explanation for the high thermal conductivity, and for the other anomalous properties observed by Allen, Peierls, and Uddin.[(2)] It is evidently possible that the turbulent motion, inevitably set up in the technical manipulation required in working with the liquid helium II, might on account of the great fluidity, not die out, even in the small capillary tubes in which the thermal conductivity was measured; such turbulence would transport heat extremely efficiently by convection.

REFERENCES

1. Burton, *Nature*, **135**, 265 (1935); Wilhelm, Misener and Clark, *Proc. Roy. Soc.*, A, **151**, 342 (1935).
2. *Nature*, **140**, 62 (1937).

E8. THE VISCOSITY OF LIQUID HELIUM*

W. H. KEESOM and G. E. MACWOOD†

Summary. An apparatus for the measurement of the viscosity of liquids and gases, interchangeably, is described. Measurements of the viscosity of liquid helium made with this apparatus are reported. The measurements were made over the temperature range from 1·2 to 4·0°K.

§ 1. *Introduction.* It was decided to continue the measurements that have been made in this laboratory on the viscosity of gases and liquefied gases which should be completed in the region of very low temperatures.

The first measurements of the viscosity of liquid helium made at Toronto[1] indicated that a more intensive investigation of this property would be of interest. Furthermore, it seemed possible that helium II might show the phenomenon of slip.

With a view to measuring the viscosity and, if possible, the slip, we decided to use a modified oscillating disc apparatus. The theory of the method has been extended by one of us[2,3] to take care of a fluid of high density and low viscosity, and exhibiting slip.

§ 2. *Apparatus.* The apparatus used is a modification of that used by van Itterbeek and Keesom[4]. It is shown in Fig. 1. The oscillation system consisted of a copper disc (*a*), 50 mm in diameter and 0·56 mm thick. This was fastened to a thin steel rod (*b*) 100 mm long and 1 mm in diameter, care being taken that it was perpendicular to the plane of the disc. The rod (*b*) was fastened with

* *Physica* **5,** 737 (1938).

† National Research Fellow, Cutting Traveling Fellow of Columbia University.

Dekothinsky in turn to a glass rod (*c*), 500 mm long and 1 mm in diameter.

Into a small opening in the rod (*c*), the suspension wire (*d*) of phosphorbronze, 250 mm long and 0·05 mm in diameter, was sealed. The other end was sealed into a heavy metal rod (*e*). The latter allowed for the raising and lowering of the suspension system.

The disc was enclosed in an open brass box (*h*) in which at the top and bottom are plane copper plates (*k*) parallel to the disc (*a*). The mechanism (*l*) allowed for changing the distance between the plates by 6 cm.

A flat mirror (*m*) was fixed to the rod (*c*) for the purpose of observing the oscillation. These were observed by means of the standard reading arrangement used in the Kamerlingh Onnes Laboratory. It consisted of a telescope and a porcelain scale, which could be read to a 0·1 mm, mounted at about 2·5 meters from the apparatus on a separate pillar. To set the system in motion, a small bar of soft iron (8 mm×5 mm) was rigidly attached to rod (*c*). Deviations were then produced by carefully bringing up a magnet.

To the large tube of german silver (*A*) a thick copper plate (*B*) was soldered. By means of the three screws (*n, o, p*), the apparatus rested on a heavy brass plate (*C*) which was fastened rigidly to a stone column standing on a pillar of the laboratory. With the aid of the screws (*n, o, p*), the apparatus could be made vertical. This done, the fourth screw (*q*) served to lock the apparatus in position.

The remainder of the apparatus shown in Fig. 1 is self-explanatory.

§ 3. *Determination of the logarithmic decrement and period.* The values of the logarithmic decrement are average values, each being calculated from about 15 complete oscillations.

The period of oscillation was determined by measuring the time of 15 oscillations with a stop-watch. The watch could be read to 0·02 sec and was checked against the Leiden Observatory clock.

§ 4. *Calibration of the apparatus.* In the case of liquid helium, as was pointed out in the paper mentioned above[(2)], the formulae

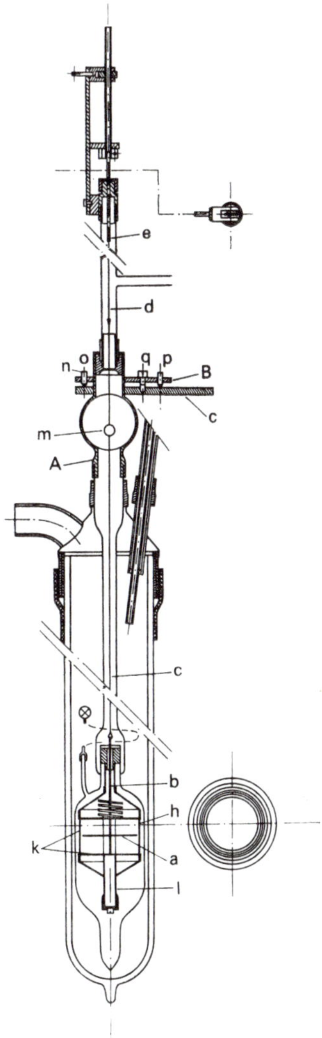

FIG. 1. Oscillating disc apparatus for measuring viscosity.

to be used for the reduction of experiments are for the viscosity and slip respectively:

$$\sqrt{\eta} = \frac{4I}{\pi R^4}\sqrt{\frac{\pi}{T\varrho}}\,(\tau^2\Delta - \tau\Delta_0)\left(1-\frac{\Delta}{2}\right), \tag{1}$$

and

$$\lambda = \frac{I}{\pi R^4}\,\frac{1}{\varrho}\,[1-(1-2\Delta)\,\tau^2 - 2\tau\Delta_0]. \tag{2}$$

where the symbols have the same meaning as in the paper referred to. However, before they may be used, $I/\pi R^4$ and Δ_0 must first be determined.

This was done by measuring the viscosity of helium gas in the liquid hydrogen range using the formula[(2)]:

$$\eta = \frac{2I}{\pi R^4}\,\frac{\bar{D}}{T}\,(\tau^2\delta - \tau\delta_0), \tag{3}$$

where $\bar{D}$ is the average separation of the disc and was measured with the aid of a cathetometer. Using the measured values of $\bar{D}$, T, τ and δ and the viscosities of helium gas as given by van Itterbeek and Keesom[(4)] and H. Kamerlingh Onnes and Weber[(5)] $I/\pi R^4$ and Δ_0 were obtained by a least square solution.

The measurements with helium gas are given below in Table I.

TABLE I. CALIBRATION OF APPARATUS

Temp. °K	Period sec.	Log. decrement × 10^4	Known viscosity μ poise	Calc. viscosity μ poise	Difference %
15·76	24·624	266·2	29·82	29·78	+0·2
17·59	24·624	286·6	32·05	32·06	−0·05
18·43	24·631	294·0	33·03	32·89	+0·5
19·14	24·624	304·0	33·82	34·01	−0·5
19·62	24·634	306·9	34·55	34·33	+0·1

TABLE II. MEASUREMENTS OF VISCOSITY OF LIQUID HELIUM

Date	Temp. °K	Period sec	Log. decre-ment × 10^5	Uncorrect-ed viscosity μ poise	Corrected viscosity μ poise
15 Dec. 1937	1·335	24·62	100·5	1·67	1·82
	1·586	24·65	114·6	2·17	2·33
	1·762	24·66	141·8	3·33	3·56
	1·907	24·67	196·0	6·36	6·75
	1·973	24·69	215·0	7·66	81·11
	2·111	24·71	269·4	12·0	12·6
	2·178	24·97	367·8	22·3	23·0
14 Jan. 1938	2·258	24·81	318·4	16·6	17·3
	2·702	24·78	404·1	27·4	28·2
21 Jan. 1938	2·08	24·71	321·4	17·1	17·8
	2·14	24·75	317·6	16·7	17·4
	2·15	24·74	294·2	14·3	14·8
11 Feb. 1938	1·731	24·60	151·2	3·80	4·10
	1·988	24·62	208·7	7·22	7·60
	2·086	24·63	256·7	10·9	11·5
	2·116	24·64	278·3	12·9	13·5
	2·145	24·65	299·4	14·9	15·5
	2·159	24·64	319·7	17·0	17·7
	2·174	24·65	336·0	18·8	19·5
	2·642	24·66	367·2	22·7	23·5
	3·81	24·66	390·6	28·0	28·7
	3·97	24·64	386·8	28·2	28·8
18 Feb. 1938	2·155	24·57	320·2	17·1	17·8
	2·156	24·61	317·9	16·8	17·5
	2·171	24·58	342·0	19·5	20·2
	2·290	24·58	371·4	18·1	18·7
	2·315	24·58	377·1	18·5	19·7
	2·934	24·69	406·1	23·7	24·0
	3·738	24·68	416·8	26·8	27·5
25 Feb. 1938	2·203	24·76	349·0	17·5	18·2
	3·069	24·79	395·3	23·2	23·9
	3·548	24·81	417·6	26·8	27·5
	3·922	24·77	412·4	27·4	28·1
	4·021	24·79	423·3	29·2	29·8
18 Mar. 1938	1·304	24·77	83·8	1·14	1·24
	1·282	24·76	90·8	1·43	1·54
	1·335	24·76	100·0	1·66	1·79
	1·379	24·77	119·5	2·43	2·61

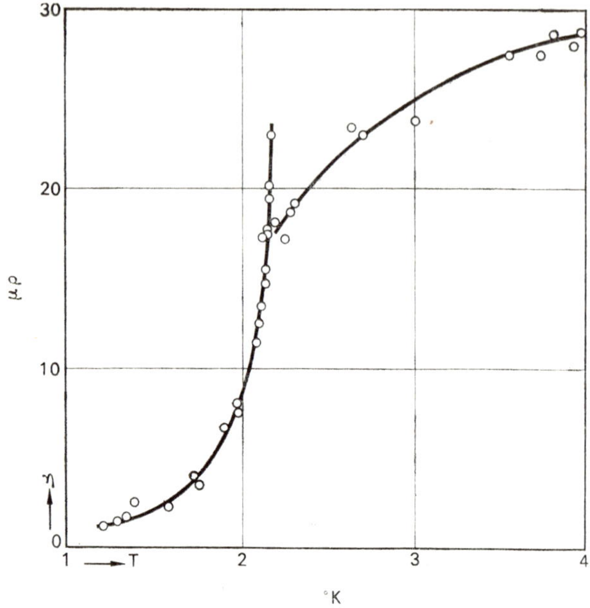

FIG. 2. Viscosity of liquid helium.

From these measurements we obtained: $\bar{D} = 0{\cdot}0605$ cm, $I/\pi R^4 = 0{\cdot}2293$ whereas Δ_0 proved to be negligible.

§ 5. *Measurements of liquid helium.* In these measurements, we observed from 10 to 15 complete oscillations for each point from which the period and decrement were obtained. On various experimental days, the disc separation was always different, varying from 1 to 1·3 cm. Considering Eq. (1), it is seen that one more datum is needed, the density. The tables given by W. H. Keesom and Miss A. P. Keesom[(6)] were used.

Table II gives a summary of the experimental results, together with the values of the viscosity according to Eq. (1). The last column gives the values of the viscosity corrected for the exterior liquid according to the calculations of one of us[(3)]. Fig. 2 gives the temperature dependence of the viscosity.

§ 6. *Discussion.* Several points of interest are immediately evident from a consideration of Fig. 2. As for most properties of liquid helium, the neighbourhood of the λ-point is abnormal for the viscosity. From the present results the course of the viscosity of He I, quite near to the λ-point, is somewhat uncertain, so that it is impossible to say whether, for the viscosity, the λ-point is merely a singular point or whether there exists a finite discontinuity there. The temperature dependence of the viscosity of both helium I and II is unusual, $\partial\eta/\partial T$ being positive, while for normal liquids it is negative.

Though the theoretical basis of Tisza's considerations[7] needs further discussion, the picture given by this author is in excellent qualitative agreement with the results given here and may be used to account for the exceedingly large differences between the values reported by Allen and Misener[8] and Kapitza[9] on the one hand, and those reported here, as was indicated by Tisza.

Furthermore, in agreement with the results of Tisza, it appears that the viscosity goes to zero as the temperature approaches zero.

Up to now, we have not mentioned slip. It had been hoped that we would be able to decide whether or not helium II possesses slip. However, our results were not sufficiently precise to decide; the most we may safely say is that, if there is a slip, it is smaller than 10^{-3}. All the calculations of the viscosity given here, therefore, were on the assumption of no slip.

Finally, it may be well to observe that the amplitude of oscillation was at all times so small that the linear velocity of the edge of the disc was never larger than 0·3 mm/sec. In Fig. 3 is shown the dependence of the amplitude as a function of time in two typical cases. On the basis of the fact, that the logarithmic decrement appears to be independent of the amplitude, we are quite sure that the motion of the fluid at all times was laminar.

§ 7. *Acknowledgements.* We wish to record our thanks to J. H. Schweers, phil. nat. drs., and P. H. Keesom, phil. nat. cand., for their help during the measurements and with the calculations.

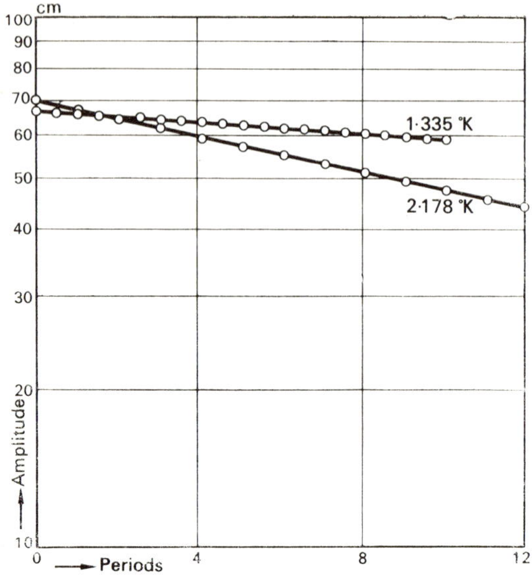

FIG. 3. Two examples of the dependence of the amplitude on the time.

Our thanks are also due to the Smithsonian Institution for a grant-in-aid.

REFERENCES

1. J. O. WILHELM, A. D. MISENER and A. R. CLARK, *Proc. Roy. Soc. London* A **151**, 342 (1935).
2. G. E. MACWOOD, Commun. Kamerlingh Onnes Lab., Leiden. Suppl. No. 84b; *Physica* **5**, 374 (1938).
3. G. E. MACWOOD, Commun. Kamerlingh Onnes Lab., Leiden. Suppl. No. 84c; *Physica* **5**, 763 (1938).
4. A. VAN ITTERBEEK and W. H. KEESOM, Commun. Kamerlingh Onnes Lab., Leiden No. 252a; *Physica,'s-Grav.* **5**, 257 (1938).
5. H. KAMERLINGH ONNES and S. WEBER, Commun. Kamerlingh Onnes Lab., Leiden No. 134b; *Proc. Kon. Akad. Amsterdam* **15**, 1396 (1913).
6. W. H. KEESOM and Miss A. P. KEESOM, Commun. Kamerlingh Onnes Lab., Leiden Suppl. No. 76b; *Physica,'s-Grav.* **1**, 128 (1933).
7. L. TISZA, *Nature, London* **141**, 913 (1938).
8. J. F. ALLEN and A. D. MISENER, *Nature, London* **141**, 75, 1938.
9. P. KAPITZA, *Nature, London* **141**, 74 (1938).

E9. TRANSFER OF HELIUM II ON GLASS*

J. G. Daunt and K. Mendelssohn

Kamerlingh Onnes[1] observed in 1922 that the surfaces of two volumes of liquid helium II, that is, helium below the λ point, in two concentric vessels adjusted themselves automatically to the same level. No further observations of this strange "distillation" phenomenon appear to have been made in Leyden or elsewhere. Rollin[2], however, found that the loss from vessels containing liquid helium II is exceptionally high, and he attributed this to some effect connected with the formation of a thin layer of helium II on the inner walls of the connecting tube. This, he suggested, might be connected with the above-mentioned phenomenon. We have lately carried out experiments with various arrangements of vessels in order to reproduce the effect observed by Kamerlingh Onnes and to investigate it systematically.

In one very convenient apparatus we utilized a small glass beaker suspended on a thin glass fibre which could be lowered into a bath of liquid helium II or raised out of it. The heat influx was reduced to a minimum, 0·5 cc of liquid at 1·5°K evaporating per hour. As it was found that light from an ordinary incandescent filament lamp was absorbed (probably on the glass wall) and disturbed the effect, our observations were carried out with a neon lamp. The following phenomena were observed:

(*a*) When the empty beaker was lowered into the liquid, it filled

* *Nature* **141,** 911 (1938).

up to the level of the bath although the rim or the beaker was everywhere above the level of the liquid.

(*b*) When the beaker was partly lifted out of the bath, the level of the liquid in the beaker fell at the same rate as it rose in (*a*), and the level of the bath rose until both had reached the same height.

(*c*) In order to establish whether the transfer took place by distillation through the gas phase or by transfer over the surface, the effect was examined of introducing "wicks" of twisted copper wire which increased the surface leading from one level to the other. It was found that this increased the rate of transfer to several times the previous rate so long as, but only so long as, the wick reached into the liquid at the higher level.

(*d*) When the beaker (without a wick) was lifted completely out of the bath, it was found that the liquid vanished at the same rate as when it was still partly dipping into it. This was accounted for

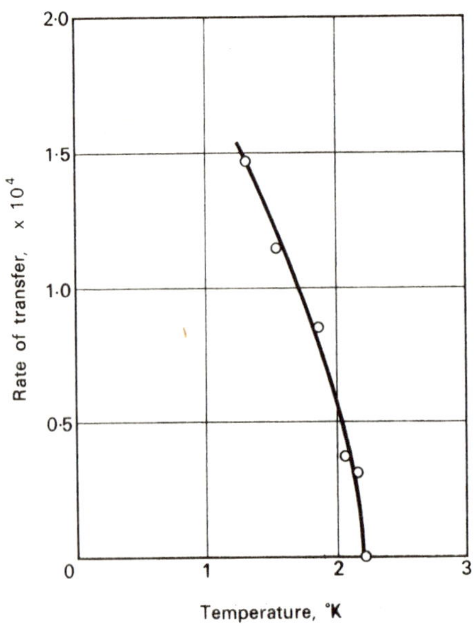

when it was observed that the liquid collected in drops at the bottom of the beaker and dripped into the bath.

(*e*) The rate of transfer did not appear to differ very greatly whether the beaker was almost full or nearly empty, the rate decreasing only by 20% when the level had dropped from within 0·5 mm. of the rim to some 20 mm. from the rim.

(*f*) As shown in the accompanying figure, the rate of transfer increased rapidly with decreasing temperature. The rate is given in cubic centimetres of liquid per second per centimetre of glass surface cutting the surface of the liquid.

(*g*) If any of these phenomena occur above the λ point, the rate is so slow that they could not be observed in our experiments.

These experiments appear to show that the surface of a solid in contact with liquid helium II is covered with a helium film through which the liquid is transferred to the lowest available level. The rate of transfer on glass (it was found to be much higher on copper) appears to be of the order which one would expect if a monatomic layer were to move with the average kinetic velocity of the molecules. But the fact that the rate of transfer increases rapidly with falling temperature seems to indicate that the phenomenon depends not so much on the molecular velocity as on the viscosity, which decreases[(3)] as the temperature falls. Whether this is to be attributed to the obvious explanation that the de Broglie wavelength increases with falling temperature is a question which must be left until further experiments have enabled a more complete theory of the two phases of the liquid state of helium to be evolved.

REFERENCES

1. Kamerlingh Onnes, H., *Comm. Leiden*, 159.
2. Rollin, B. V., Actes 7. Cong. int. du Froid, **1**, 187; Prof. Simon informs us that a full account will be published shortly.
3. Allen, J. F., and Misener, A. D., *Nature*, **141**, 75 (1938).

E10. THE STUDY OF HEAT TRANSFER IN HELIUM II*

P. L. KAPITZA

Summary. Keesom and Keesom showed that liquid helium II in capillary tubes possesses an unusually large heat conductivity which, by analogy to superconductivity in metals, they named "superheat-conductivity".

In opposition to this view the author put forward a hypothesis which held that this abnormal heat conductivity was not due to some exceptional thermal property of helium II but to heat transferred by convection currents whose presence can be anticipated owing to the exceptionally high fluidity of liquid helium II, and the author suggests that it should be named "superfluidity".

That the heat conductivity is due to these convection currents is established by the experiments described. In the first instance, when the heat conductivity was studied in capillaries it is shown that the heat conductivity of helium II in a capillary may be found to be so great, and the amount of heat transferred so large, that the values obtained cannot be explained by any of the usual mechanisms of heat conductivity without contradicting the fundamental views held on heat phenomena.

Then it is shown that the heat conductivity of helium II in capillaries was a hundred times less if the liquid was set in motion. Further experiments showed the character of the flow of the helium II and made it possible to measure its velocity. There appeared to be two currents; one flowing along the walls of the capillary and the other flowing in the opposite direction down the centre.

A hypothesis was put forward which held that the heat transfer of helium II is produced by these convection currents and is due to a difference of the heat content of the two currents of helium flowing in opposite directions. The author assumes that the helium II flowing in very thin films on the walls has a different heat function owing to the molecular force of the surface. The author showed that this suggestion is sufficient to explain the abnormal heat conductivity of helium II. In this way the author came to the conclusion that the heat conductivity of helium II is due only to the high velocity of

* *J. Phys. U.S.S.R.* **4,** 181 (1941).

the flow of the helium in the film which is possible owing to its "superfluidity".

Further, the author's views were confirmed by experiments in which the heat transfer of a body freely hung in helium was studied and in this case no abnormal heat conductivity was observed.

1. INTRODUCTION

Liquid helium below the λ-point acquires a number of most peculiar physical properties which do not occur in other liquids. One of the most interesting is an exceptionally high heat transfer in capillaries. This discovery was made by Keesom and Miss Keesom[1] who even described this exceptionally high heat transfer by an analogy to superconductivity in metals as a "superheatconducting" phenomena.

A number of researches have confirmed this interesting result but the nature of the heat transfer in helium II was left an open question. The authors of the discovery, including their recent work[2], have taken the view that this phenomenon is an anomalous heat conductivity but of the ordinary type. Contrary to this view, three years ago, we put forward the suggestion that it may be possible to explain this anomalously big heat transfer in helium II as the result of a very efficient convection mechanism. The existence of intense convection currents was considered possible owing to the immeasurably small viscosity of helium II which we had established by studying its flow through a narrow slit[3]. Therefore, we put forward the view that helium II must be regarded as a "*superfluid*" rather than as a "superheat-conducting" liquid.

After the publication of our work the question concerning the absence of viscous motion in helium was repeatedly discussed and a number of experiments were entered upon. Divergent views exist upon the subject. Some of the research workers, Burton[4] and Giauque[5] for example, believe that in our work we did not pay attention to the "creeping" of the helium up the sides of the vessel which would account for the smallness of the observed value of the viscosity. This assertion is not justified as from the experiments of Daunt and Mendelssohn[6] it appears that the amount of helium creeping up the walls is so small that it cannot affect the results of our experiments materially. To prove this we also made control experiments in which the slit in our viscosity meter was sealed with oil and in this way the creeping out effect could be checked and it was found that the influence of "creeping" is less than the limits of experimental error.

On the contrary, it seems to us that the creeping phenomenon in helium II is itself one of the best phenomena for demonstrating the absence of viscosity. A simple calculation from the data of the experiments of Daunt and Mendelssohn and also Kikoin and Lasarev[7] shows that the rapid flow observed in a helium film 10^{-5} to 10^{-6} cm thick is only possible if the viscosity is even smaller than the already small limit (10^{-9}) established by our experiments.

In the following work we studied the mechanism of heat transfer. It seems to us that the results of this work establish a dynamic picture of the heat transfer in helium II and confirm our original view. The picture at which we arrive, though not so simple as we at first thought, leads us to the conclusion that the mechanism of heat transfer can be explained by the property of "superfluidity" of helium II.

It may be mentioned that this conclusion agrees with the general views of heat conductivity and viscosity as accepted in physics. In fact, the existence at one and the same time of a high heat transfer and lack of viscosity would be a very unexpected property of matter. If we take, for instance, gases as an example, both the heat transfer and the viscosity are determined by the efficiency of the mechanism of transfer of momentum from one atom to another. It is natural to expect that the bigger the heat conductivity the greater is the viscosity.

On the other hand, in any experiments made to measure heat transfer and viscosity all the main disturbing effects such as turbulent motion, convection currents, etc., will only increase the values obtained. The values obtained from experiments on heat transfer will always be overestimated and the fluidity will always be underestimated.

Below we give the experimental results of the last two years' work in the order in which it was done with the conclusions to which they lead us.

2. HEAT TRANSFER IN CAPILLARIES

The problem

At the beginning of our research work we set ourselves the problem of determining the velocity of circulation in helium necessary to account for its abnormal heat transfer by convection currents. Let us take first of all the case of classical heat conductivity in a cylindrical body of section S and length l. Let a current of heat equivalent to w Watts flow along the axis giving at the ends of the cylinder a temperature difference of ΔT, then

$$w = k\frac{S}{l}\Delta T; \tag{1}$$

hence k is the heat conductivity,*

$$k = \frac{w}{S}\frac{\Delta T}{l} \text{ W/degree cm.} \tag{2}$$

In cases when we have a heat transfer produced by a convection mechanism the amount of heat carried will be determined by the product of the velocity of the convection current v, the specific capacity c and mean density ϱ.

* In this paper, when calculating the heat conductivity we shall use Watts instead of calories per sec. This quantity is evidently 4·18 times greater than the usual value.

Hence we have

$$w = 4 \cdot 18 c\varrho v S' \Delta T, \tag{3}$$

where the numerical coefficient is the heat equivalent of the electrical power. The part of the cross-section of a tube through which the convection currents flow must obviously be smaller than its full cross-section S, namely:

$$S' < S. \tag{4}$$

Therefore, if this is the only way in which the heat transfer takes place we get for the velocity of the convection v and the lower limit of v':

$$v' = \frac{w}{4 \cdot 18 S} \cdot \frac{1}{c\varrho \Delta T} < v. \tag{5}$$

To determine the value of the convection velocity of the liquid necessary to account for the observed large values of heat transfer in helium II, it is seen from formula (5) that there is no need to perform experiments with long capillaries as the length does not enter into the definition of v'. As is known from the experiments of Keesom and his collaborators one gets the largest value for the heat transfer only with small loads and correspondingly smaller temperature drops. Therefore, for finding the largest possible values of the velocity of convection v' it is most important to have a sensitive method of determining the difference of temperature at the ends of the capillary.

The apparatus

Figure 1 shows one of the suitable apparatuses for our experiments. The capillary *1* in which the heat transfer of the helium II was measured, was 4 cm long with an inside diameter of 0·61 mm and an outside diameter of 1·21 mm. A glass rod of diameter 0·47 mm could be inserted into the capillary and in this way the section of the capillary was reduced without changing any of the other conditions of the experiment. As is seen from the drawing (Fig. 1), the thermometer and the heater were placed in the bulb *3* which was designed in such a way that the free space filled with helium II was made as small as possible. This space was usually

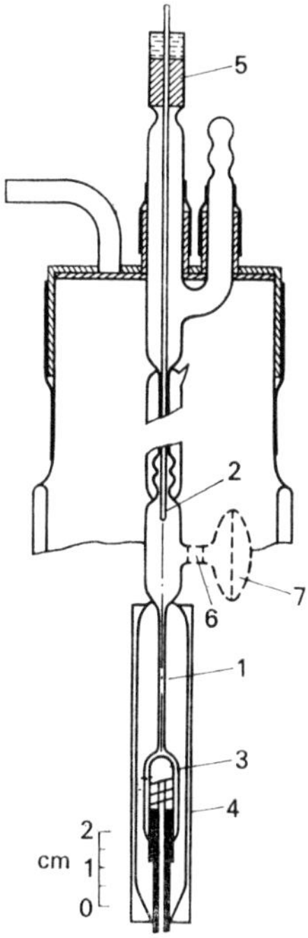

FIG. 1. The apparatus for measuring the heat conductivity in capillaries with an arrangement for reducing the cross-section.

about 1–2 cm³. The heater was a thin constantan wire with a resistance of about 30 ohms. The temperature was measured with a bronze thermometer made of wire 300 cm long and 0·1 mm in diameter. To the thermometer and heater were attached potential and power leads. The capillary and the bulb were surrounded by a vacuum insulation, and to prevent heating from radiation a copper screen *4* was provided. The whole apparatus was attached to a glass tube with a vacuum insulation and fixed into a Dewar vessel. The rod *2* could be lowered or raised during the experiment through the rubber stopper *5* sealed with glycerine. The capillary *1* and the bulb *3* were filled with helium through the opening *6* and to this, as will be described in the next paragraph, a copper ball

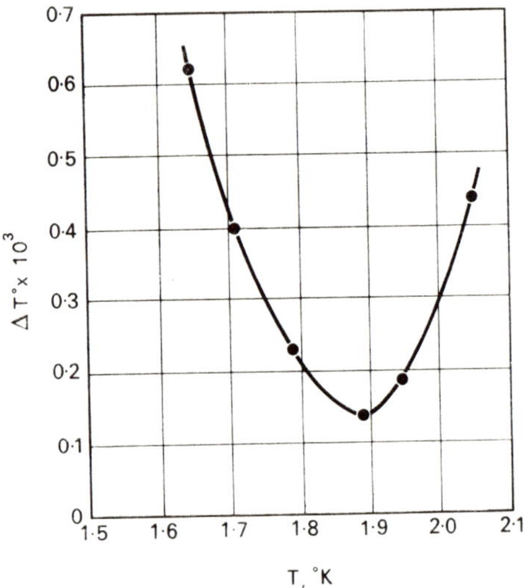

FIG. 2. The temperature drop along a capillary 4 cm long, with a diameter of 0·61 mm, at a load of 0·45 W/cm².

7 pressed out of thin copper sheet could be soldered. In the experiments described in this section this ball was absent.

The differences of the potential on the bronze thermometer at a current of 1 mA was measured by a potentiometer. The measurements of the small temperature difference along the capillary produced, when the heater was set in operation, were made simply by the deflections of the galvanometer which were recorded on a moving tape of photographic paper. Examples of the records so obtained are shown in Fig. 5.* The method of photographic record-

* *Note by Editor:* This figure is part of the text which is not reprinted.

give no abnormally high values for the heat conductivity of helium II and it is shown that most probably its value does not exceed that of helium I.*

REFERENCES

1. W. H. KEESOM and MISS KEESOM, *Physica*, **3**, 359 (1936); **5**, 218 (1938).
2. W. H. KEESOM and S. F. SARIS, *Physica*, **7**, 241 (1940).
3. P. KAPITZA, *C. R. Acad. Sci. URSS*, **18**, 21 (1938); *Nature*, **141**, 74 (1937); *Collected Papers*, ch. **34**, p. 505.
4. E. F. BURTON, *Nature*, **142**, 72 (1938).
5. W. F. GIAUQUE, J. W. STOUT and R. E. BARIEAU, *Phys. Rev.*, **54**, 147 (1938); *J. Am. Chem. Soc.*, **61**, 654 (1939).
6. J. G. DAUNT and K. MENDELSSOHN, *Proc. Roy. Soc.*, **A 170**, 439 (1939).
7. I. K. KIKOIN and B. G. LASAREV, *Nature*, **142**, 289 (1938).
8. W. H. KEESOM and MISS A. P. KEESOM, *Physica*, **2**, 557 (1935).
9. P. DEBYE, *Vorträge über die kinetische Theorie der Materie der Elektrizität*, Teubner, p. 3 (1914).
10. J. F. FINDLAY, A. PITT, H. GRAYSON SMITH and J. G. WILHELM, *Phys. Rev.*, **54**, 506 (1938).
11. J. F. ALLEN and E. GANZ, *Proc. Roy. Soc.*, **171**, 242 (1939).
12. J. F. ALLEN and H. JONES, *Nature*, **141**, 243 (1938).
13. P. G. STRELKOV, *J. Phys. USSR*, **3**, 53 (1940).
14. A. VAN ITTERBEEK, W. VAN DINGENEN and V. BORGHS, *Physica*, **6**, 951 (1939).

* *(Note by Editor)*. We may quote here how Kapitza himself saw the situation three years later:

"As I said before, I explained this phenomenon originally by the filling of the vessel by helium, through the flow of helium in the opposite direction in a thin layer. I assumed also that the energy state of helium II is that a thin layer differed from the energy state of free helium II, and could thus explain the apparent large thermal conductivity of helium. I could also approximately evaluate the thickness of this layer by assuming that the flow velocity of the helium in it would not be excessively large. I also—as mentioned earlier—tried in these experiments to observe experimentally the thickness of this layer. To do that, I forced helium to flow in a very thin layer. I gradually reached a helium layer thickness of 0·00014 mm, but the experiment showed that the character of all phenomena remained unchanged. We needed thus to reconsider this explanation and this led to completely new ideas about the nature of hydrodynamics phenomena in helium II. The first sketch of these ideas is due to Tisza but our scientific colleague L. Landau was the person who worked them out, gave them a theoretical foundation, and produced a hydrodynamic theory of the various phenomena." (Kapitza, *Uspekhi (USSR)* **26** 133 (1944); *Collected Papers*, Vol. 2.)

E11. A DIRECT OBSERVATION OF TWO KINDS OF MOTION IN HELIUM II*

E. Andronikashvili

Summary. The existence of two kinds of flow in helium II (viscous and potential) is confirmed by experiments. As it is known, these flows may be characterized by introducing the concept of two effective densities of He II: a "normal" density ϱ_n and a "superfluid" one ϱ_s.

The temperature dependence of the ratio ϱ_n/ϱ_s is investigated.

1. The phenomenon of superfluidity discovered by P. Kapitza(1) is, together with the superconductivity, one of the most principal problems of low-temperature physics. London(2) and Tisza(3) introduced for the explanation of this phenomenon the conception of the division of helium into "superfluid" and "normal" components, each of which is represented by a certain number of atoms, depending upon temperature, occupying the zero or different from zero "energy levels".

However, such a model of liquid helium gives only a qualitative picture of the phenomenon—the flowing of helium through capillaries—and is quite unable to account for the peculiar hydrodynamical properties of helium II.

Further investigation of superfluidity of helium II performed by Kapitza(4) gave an extensive experimental material which enabled L. Landau(5) to work out a rigorous quantitative theory of this phenomenon, containing also the conception of superfluid and normal masses of helium II. But contrary to the hypothesis of London and Tisza in Landau's theory there corresponds to every

* *J. Phys. (U.S.S.R.)* **10**, 201 (1946).

component of helium II not a definite number of atoms, but a definite kind of motion: the potential flow of an ideal fluid or the flow of viscous fluid possessing friction.

The superfluid mass, having no friction, is not carried along by bodies moving in the fluid (if the fluid is not stirred up). On the contrary, the normal mass is carried along under any conditions by bodies moving within helium II, due to the ordinary friction.

The ratio between the both masses at different temperatures was calculated by Landau. Here we note only that it varies from $\varrho_n/(\varrho_n+\varrho_s)=1$; $\varrho_s/(\varrho_n+\varrho_s)=0$ at the λ-point to $\varrho_n/(\varrho_n+\varrho_s)=0$; $\varrho_s/(\varrho_n+\varrho_s)=1$ at 0°K. ϱ_n and ϱ_s being the normal and superfluid densities. In accordance with this the author of the theory proposed an ingenious experiment which could directly confirm the existence of the superfluid and the normal masses and also allowed to establish experimentally the temperature dependence of their ratio.

The idea of the experiment is the following: the moment of inertia of a vessel, containing helium II, will vary from a maximum value at the λ-point (the moment of inertia of the vessel itself plus the moment of inertia of the whole quantity of helium II contained in it), to the minimum value at 0°K, when all helium will be at rest.

The experimental study of the conception of normal and superfluid masses which is fundamental for the theory of superfluidity is the object of the present paper.

2. It is a difficult problem, because of the low viscosity of the liquid, to convert helium into the state of rotation (without stirring it up) during the period of time consistent with the conditions of the experiment. For helium I the viscosity equals approximately $2\cdot10^{-5}$ poises. In order to solve this problem, the distance between the rotating parts of the apparatus must be of the order of the penetration depth λ, which is defined, as is known, by the expression:

$$\lambda=\sqrt{2\eta/\omega\varrho},$$

where η is the viscosity of the liquid, ϱ—its density and ω—the circular frequency.

In the case of the torsional vibrations with a frequency $\omega = 2\pi$ sec^{-1}, the penetration depth in helium 1 equals approximately only 0·1 mm. Hence the distance between the rotating surfaces must be of the order of 0·2 mm. It is necessary to keep in mind that the mass and the moment of inertia of the whole apparatus must be as small as possible, because in the opposite case the

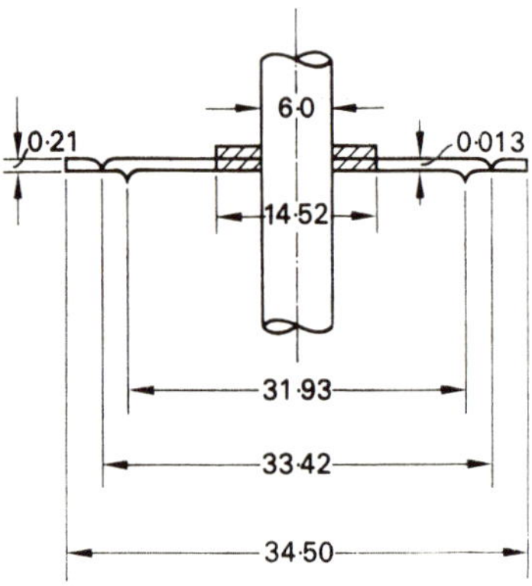

FIG. 1

variations of the moment of inertia of helium II to be measured will be quite insignificant.

The scheme of the apparatus adapted for this purpose was given in principle by P. Kapitza. A pile of aluminium sheets, cut from aluminium foil approximately 0·013 mm thick, interlaid by tissue paper, was set on an axis and was turned in a lathe up to the diameter 34·5 mm.

In order to give to these sheets the necessary rigidity, each of them was pressed between a matrix and a stamp and in this way a narrow rib 0·21 mm high was formed on the surface of the sheet. Two kinds of matrix and stamps were used; the first pair formed a rib 33·42 mm in diameter and 0·39 mm broad on one part of the sheets; the remaining part was pressed between the other pair and had a rib 31·93 mm in diameter and 0·58 mm broad (Fig. 1).

Hundred sheets treated in this way were set in turn on a hollow aluminium axis 6 mm in diameter resting on each other by ribs, having equal heights, but shifted along the diameter with respect to one another.

In the middle the sheets were supported by aluminium washers 0·21 mm thick and 14·52 mm in diameter, they were set on the

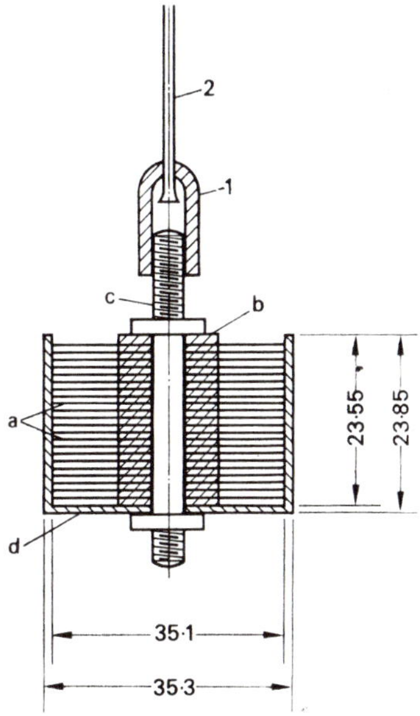

FIG. 2

axis in turn with the sheets. In this manner the necessary distance and the parallelism between the surfaces was obtained—the parallelism brings to a minimum the stirring up of helium during the rotation of the apparatus. Hundred sheets *a* and hundred washers *b* gathered into a pile were put into an aluminium vessel *d*, which had the following dimensions: 35·1 mm inner diameter, 23·85 mm height, 0·1 mm wall thickness. The bottom thickness was 0·3 mm.

The vessel was rigidly fastened to the common with the pile

c axis. The eccentricity of the assemblage equalled 0·2 mm. The weight of the apparatus equalled 16·10 g (Fig. 2).

By means of the aluminium cap *1* screwed on the end of the axis, to which a rectilinear glass rod *2*, 500 mm long and 2 mm in di-

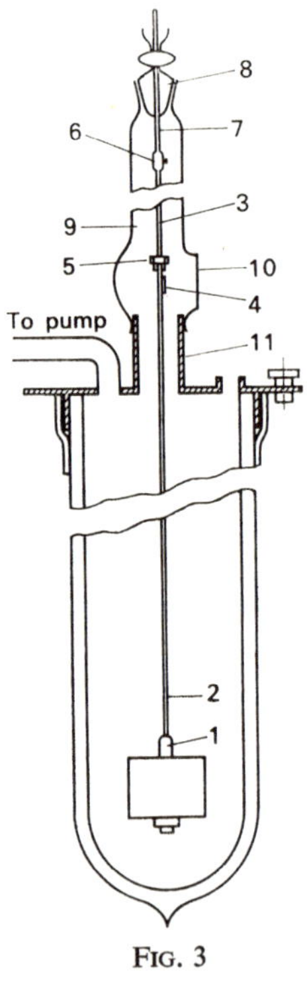

FIG. 3

ameter, was connected, the apparatus was suspended on a thin phosphorus-bronze wire *3* 40 μ in diameter and 100–120 mm long. In the upper part of the rod a small mirror *4* was sealed on, and at the end of the rod a steel clamp *5* was fitted by means of de Khotinsky cement (the end of the wire was inserted into this clamp); the clamp served simultaneously for the purpose of rotation of the apparatus which was performed by bringing a magnet to the clamp. The other end of the wire was fastened by means of the clamp *6* to the metallic rod *7*, that passed through the picein seal within the cone of the glass ground joint *8*.

The cone of the glass ground joint that supported the entire suspended system rested on a cylindrical glass tube *9* (with a flat window *10*) which was put on a stebrite tube *11* sealed into the cap of the helium Dewar vessel, to which it was fastened by means of picein.

Thus, during the whole time of experiment the wire was held at room temperature (Fig. 3).

The moment of inertia of the suspended system (the apparatus, glass rod, clamp and the arretire) amounted to 11·70 g·cm^2 and was determined during the preliminary experiments with an accuracy up to ±0·3%.

Under these conditions the moment of inertia of helium I amounted to 50% of the moment of inertia of the apparatus.

The beam of light, reflected by the mirror, fell on a scale distant 1·5 m from the mirror. The width of the light spot did not permit to read the scale divisions with an accuracy exceeding ±0·5 mm. By means of this arrangement it was possible to determine the damping decrement, the knowledge of which was necessary, for example, for the purpose of control of the laminar character of the movement.

Naturally it was our aim to perform the experiment under such conditions when the entire part of the "normal" fluid carried along by friction would move during the rotation of the apparatus as a whole. Therefore, we have chosen the period of vibrations as great as possible of the order of 25 sec.

The period was determined by measuring 10 full vibrations by means of a stop watch, the readings of which during a long period of time were compared with the second signal of the central clock of the Institute for Physical Problems.

The measurement of the period was repeated for many times.

The value of the period was obtained with an accuracy up to $\pm 0{\cdot}08\%$.

3. In the case when the penetration depth is large, compared with the distance between the rotating discs, as it took place in our experiments ($2\,\lambda \approx 4a$, a being the distance between the discs) we are able to calculate the moment of inertia or, what is the same, the mass of the rotating liquid, starting from quite elementary considerations.

In the geometrical conditions of the apparatus described, we can write the following simple expression for the moment of inertia of the entire system:

$$J = J_v + J_{He} = J_v + \int_r^R 2\pi \varrho_n a N r^3\, dr = J_v + \frac{\pi}{2} a N \varrho_n (R^4 - r^4) = \frac{K}{\omega^2},$$

where J_v—moment of inertia of the vessel, a—distance between the discs, N—their number, R—radius of the vessel, r—radius of the axis (in our case, radius of the intermediate washers), ϱ_n—density of the "normal" component of helium, K—torque of the suspension per unit angle and ω—frequency of the vibrations.

Then for the density which takes part in the rotation

$$\varrho_n = \frac{2(K - J_v\omega^2)}{\pi a N (R^4 - r^4)\omega^2} = \frac{2J_v(\omega_0^2 - \omega^2)}{\pi a N (R^4 - r^4)\omega^2};$$

here ω_0—the vibration frequency of the empty vessel.

The exact solution of the hydrodynamical problem (the solution of the Navier–Stokes equation with the boundary condition, corresponding to the conditions of our experiment) was obtained by K. A. Tumanov, and showed that the correction due to the incomplete setting of the liquid into rotation does not exceed 1%. As the inaccuracy in the measurement is larger, we did not introduce it into the calculation of ϱ_n.

But the expression for ϱ_n, given above, does not take into account the fact that a thin layer of liquid is carried along by the outer wall of the rotating vessel. This correction, which amounts to several per cent, can be accounted for by introducing a certain coefficient. Then the expression for the density of "normal" component will have the form

$$\varrho_n = \frac{2J_v(\omega_0^2-\omega^2)}{\pi Na\omega^2(R^4-r^4)} \quad \frac{1}{1+\lambda\dfrac{1+(2L/R)}{aN}},$$

where λ is the penetration depth and L—the height of the vessel.

From the measurements of Keesom and McWood[6], as well as from our preliminary experiences on the viscosity of liquid helium it follows, that the penetration depth, in helium II, under the conditions of the experiment described, equals $3{\cdot}8 \cdot 10^{-2}$ cm. Here the ratio of the kinetical viscosity ν (depending on temperature) to the frequency ω (altering with the lowering of the temperature, due to the decrease of the rotating part of helium) remains practically constant (this is true in any case for the temperature interval from λ-point to 1·85°K).

The weak temperature dependence of the penetration depth does not exceed 2% and produces the instability of the correction coefficient which at any rate is less than 0·2%.

The small jump of the penetration depth, observed at λ-point during the transition from helium I to helium II, alters the value of the correction coefficient also quite insignificantly, less than by 1% and, therefore, is not taken into account also. Thus, for all temperatures the correction introduced remains constant within the limits of experimental errors.

As the ordinary (total) density of helium below the λ-point does not change (with an accuracy up to 0·5%), then, observing the variation of the period of oscillations when the temperature is lowered, we are able to determine in a very direct way the ratio between the "normal" and the "superfluid" components.

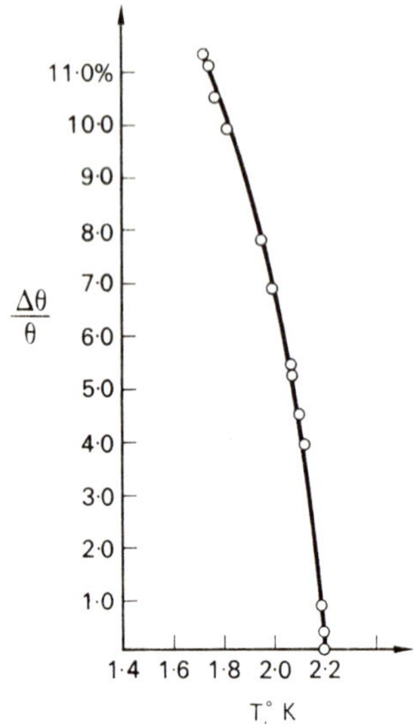

FIG. 4

4. The dependence of the period of the vibrations on the temperature follows from Fig. 4. The curve is plotted according to the data, obtained with a period equal to 25 sec.

Assuming, as it was shown above, that the coefficient

$$k = \frac{1}{1+\lambda\,\dfrac{1+(2L/R)}{aN}}$$

does not depend on temperature, and measuring the periods of vibrations of the vessel filled with helium I, $\theta_{\rm I}$, filled with helium II, $\theta_{\rm II}$, and of the empty vessel, θ_0, we obtain for the ratio of the "normal" and the total densities the following expression:

$$\frac{\varrho_n}{\varrho} = \frac{\theta_{\rm II}^2-\theta_0^2}{\theta_{\rm I}^2-\theta_0^2}.$$

Determining the period by measuring the duration of 10 full

vibrations which makes about 250 sec. and allowing here an error of the order of 0·2 sec. or 0·08%, we are able to compute ϱ_n/ϱ with a sufficient accuracy. Inserting into this formula the values of the periods from Table 1 and determining the relative error of the measurements of this quantity we find that this error

$$\frac{\Delta(\varrho_n/\varrho)}{\varrho_n/\varrho},$$

depending on the temperature, is 1·9% in the λ-point and reaches 3·9% at $T = 1{\cdot}76°K$.

The determination of the total density of helium I near λ-point performed by the same method (the measurement of the moment of inertia of the vessel filled with helium) gave the value equal to 0·1465 g/cm³, which coincides with its true value.

Table 1 and Fig. 5 give the temperature dependence of ϱ_n/ϱ.

TABLE 1

T (°K)	θ_0	θ_I and θ_{II}	ϱ_n	ϱ_n/ϱ
2·25	22·58	26·77	0·1465	1·00
2·21	22·76	26·97	0·146	1·00
2·18	22·58	26·71	0·143	0·983
2·155	22·76	26·60	0·132	0·906
2·09	22·76	25·85	0·105	0·718
2·06	22·58	25·37	0·0945	0·646
2·00	22·76	25·21	0·0820	0·561
1·95	22·58	24·68	0·0700	0·479
1·88	22·76	24·44	0·0554	0·379
1·76	22·58	23·93	0·0441	0·302

As can be seen from the curve in Fig. 5, the measured values of ϱ_n/ϱ (circles) agree well with the values obtained in an indirect way from the investigation of the velocity of the second sound in helium II, performed by Peshkov (these data, which he kindly

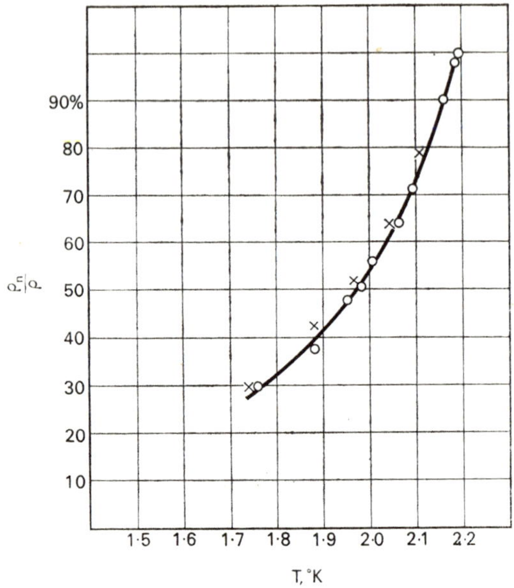

FIG. 5

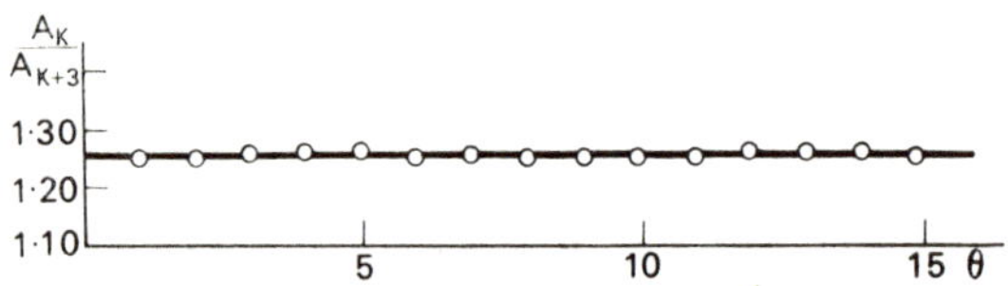

FIG. 6

communicated to me, are given by crosses). The theoretical values calculated in the paper of Landau[5] also agree quite closely with the experimental curve.

The curve in Fig. 6 gives the typical dependence of the logarithmic damping decrement on time. It follows from the curve that there is no break of laminar flow due to the stirring up of the liquid or to the appearance of the turbulence.

It is a pleasant duty to express my deep gratitude to Prof.

P. Kapitza for numerous very important advices and for permitting me to use the facilities necessary for the performance of the present investigation.

I am also very grateful to Prof. L. Landau, V. Peshkov and Prof. A. Shalnikov for the valuable discussions.

REFERENCES

1. P. KAPITZA, *Nature*, **141,** 74 (1938).
2. F. LONDON, *Nature*, **141,** 643 (1938).
3. L. TISZA, *Nature*, **141,** 913 (1938).
4. P. KAPITZA, *Journ. of Phys.* **4,** 181 (1940); **5,** 59 (1941).
5. L. LANDAU, *Journ. of Phys.* **5,** 71 (1941).
6. W. KEESOM and G. MCWOOD, *Physica*, **5,** 737 (1938).
7. V. PESHKOV, *Journ. of Phys.* (in press).

E12. DETERMINATION OF THE VELOCITY OF PROPAGATION OF THE SECOND SOUND IN HELIUM II*

V. PESHKOV

Summary. A detailed description is given of the experiments on the observation of the second sound in helium II. The velocity of the second sound was measured with an accuracy up to 0·3 per cent in the temperature region 1·36–2·19°K, and was used for calculation of the ratio of the superfluid and normal "parts" of the liquid according to Landau's theory.

INTRODUCTION

It is well known that helium is the only substance which exists in the liquid state down to absolute zero. As the temperature is lowered helium I which is a liquid with the usual properties transforms at the λ-point (2·19°K) into helium II which is a liquid with some very peculiar properties. Helium II has attracted the attention of a large number of investigators.

Keesom and Miss Keesom[(1)] discovered the very high thermal conductivity of helium II in capillary tubes and explained it by the superconductivity of this liquid.

However, the detailed and careful experiments of Kapitza[(2–4)] proved that helium is not thermally superconductive but primarily a superfluid, superfluidity being the fundamental property of helium II. Thus the high thermal conduction of helium II is really due to the peculiar convection currents which are found only in helium II.

* *J. Phys. (U.S.S.R.)* **10,** 389 (1946).

In an attempt to explain theoretically the extraordinary properties of helium II London[5] and Tisza[6] considered it as a gas obeying the Bose–Einstein statistics and consisting of a mixture of atoms, these in the excited state possessing the usual viscosity properties while the second sort of atoms which were assumed to exist in the degenerate state with zero energy were assumed to possess the property of superfluidity. However, the theoretical speculations of London and Tisza are very artificial and unconvincing.

Landau[7,8] studied the phenomenon of superfluidity on the basis of the quantum mechanics and proposed the theory of a quantum liquid—helium II. This theory not only explained the main experimental results of Kapitza but also predicted the existence of a second sound in helium II. In his solution of the problem of propagation of sound in helium II Landau obtained two wave equations with different velocities instead of the sound equation and the usual equation for rapidly damped thermal waves. One of these velocities corresponds to that of the usual sound (which according to the measurements of Findlay[9] and others is of the order of 240 m/sec) while the second depends strongly on the temperature and is of the order 25 m/sec at 1·33°K.

An attempt to detect the second sound by the beats in standing waves radiated by oscillating piezoquartz was undertaken in the Institute for Physical Problems by Shalnikov and Sokolov before the war, but without success.

On analysing the cause of the negative result of this experiment, Lifshitz[10] calculated that the oscillating plate radiates second sound a million times less intensively than the ordinary one (first sound). On the other hand, a plate with periodically varying temperature radiates second sound about $5 \cdot 10^3$ times more intensively than the first sound. He pointed out also that the observation of the oscillations of the temperature is more favourable than that of the pressure.

After these calculations the cause of the failure of the first experiments was clear and Kapitza suggested that I make a new attempt to detect the second sound in helium II.

On the basis of Landau's theory one may mentally draw the following picture of helium II. At absolute zero it is a liquid with no thermal motion. As the temperature is raised thermal excitation begins to appear not continuously, however, but as quanta. These thermal quanta exist in the form of phonons or quanta of sound, and of rotons—which are stable vortex formations. Definite portions of energy are required for the production of both types of thermal quanta.

If helium flows along a wall slowly its interaction with the wall may be too weak to excite thermal quanta. In this case the energy of the translatory motion cannot transform into thermal energy and then appears the fundamental property of helium II—its superfluidity.

It is clear that beginning with some critical velocity the interaction of helium with the wall increases up to a limit at which thermal quanta begin to appear and helium loses its property of superfluidity. This velocity was observed by Kapitza.[(4)]

The thermal quanta in helium II behave similarly to a gas. On collision with the wall they transfer a part of their momentum to it, for this reason even if a plate moves in helium II with a velocity considerably below the critical value it is subjected to viscous friction due to the interaction with the thermal quanta. Such viscous forces were really detected by Keesom and Macwood[(11)] for an oscillating disc in helium II.

If helium II is forced through thin slits as in Kapitza's experiments[(4)], then the thermal quanta will move through the slit with a very small velocity while helium as a whole will move through it without experiencing any viscous forces (if the velocity does not exceed the critical value) due to its superfluidity. The slit thus filters out the thermal quanta and the outflowing helium has a temperature close to absolute zero. This fact was noted by Kapitza and was suggested by him as a method of obtaining superlow temperatures.

Landau showed that phenomenologically helium II may be considered as consisting of two parts, of a superfluid and a normal constituent, the former having a density ϱ_s and the latter—a density ϱ_n, where $\varrho_s+\varrho_n = \varrho$ is the density of helium. The velocity of helium is likewise given by the formula $\varrho v = \varrho_s v_s + \varrho_n v_n$, where v is the velocity of helium as a whole and v_s and v_n are respectively the velocities of the superfluid and normal components.

The expediency of such a division may be illustrated by the following experiment. If a column of closely arranged discs is made to oscillate in helium II, then the thermal quanta, i.e., the normal component of helium II, will be completely carried along while the superfluid component will remain at rest. Thus by measuring the change in the moment of inertia one may determine what fraction of helium mass is connected with the thermal quanta and what fraction is superfluid. Such experiments were carried out by Andronikashvili[(12)] and the results are in good agreement with the theory.

Heat conduction in helium II is also due to the peculiar convection currents. The normal component of helium II moves in one direction and carries with it all the thermal energy while the superfluid component flows in the opposite direction. This mode of heat conduction is many times more effective than that of the best heat conductors.

In order to clarify this motion of helium within itself the following analogy from hydrodynamics may be cited. As is well known, any running wave possesses a certain momentum $p = W/u$, where W is the energy carried by the wave per unit time and u is the velocity of propagation of the wave. Thus, if in a liquid of mass M contained in a closed volume sound waves travel from one wall to the other, it is necessary, in order that the liquid as a whole remains at rest, to assume that it moves with a velocity $v = W/Mu^2$ in a direction opposite to that of the sound. This phenomenon may be described in a different manner. Suppose that only a part of the liquid mass

$$M_n = \frac{p}{u} = \frac{W}{u^2}$$

is related to the sound motion. Then the sum of the remaining part M_s and M_n will equal the total mass, i.e., $M_s + M_n = M$. But as the liquid is at rest $M_s v_s + M_n v_n = 0$ which means that within the liquid two parts of it move in opposite directions.

A similar state of affairs exists in helium II but instead of a regular sound motion a chaotic motion of phonons and rotons is connected with the excited part. At the λ-point the complete mass of helium begins to participate in the thermal motion and helium loses its remarkable properties, transforming into an ordinary liquid. Finally, one other singular phenomenon might be expected in helium II. Thermal quanta are similar to gas particles not only in respect to their collisions with the walls but also in the sense that they collide and transfer momenta to each other. This analogy is so fargoing that there even exist "sound" vibrations of a gas consisting of thermal quanta. This is exactly the second sound which is investigated in the present paper.

EXPERIMENTAL

As the second sound in helium II is chiefly a kind of thermal vibrations, in order to radiate and detect it we used a thermal method. For the radiator we used a heater which imitated a plane capable to radiate rapidly oscillating thermal flow. A resistance thermometer consisting of a very thin wire of phosphor-bronze was used as the indicator. Wires of various diameters obtained by drawing without tempering were prepared for the thermometer. The dependence of the resistance on temperature for wires of phosphor-bronze obtained from a single sample by this method is illustrated in Fig. 1. Relative values are given in the curve, the resistance at room temperature being accepted as unity.

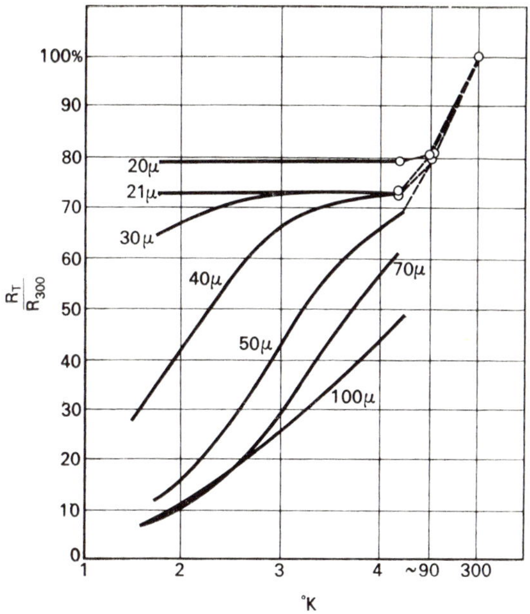

FIG. 1. Dependence of the resistance of phosphor-bronze wires of different diameters upon temperature.

The idea of the experiment is the following. If an alternating current of frequency $\omega/2$ is passed through a low inertia heater the amount of released heat will equal

$$W = RI^2 = RI_0^2 \cos^2 \frac{\omega t}{2} = \frac{RI_0^2}{2}(1+\cos \omega t),$$

i.e. besides a constant component a variable component of frequency ω appears. In the usual liquids ordinary rapidly damping heat waves are propagated from such heater. Due to the singular properties of helium II, according to Landau's theory, such thermal waves will not be damped in this liquid but will be propagated as the second sound accompanied by a periodically varying temperature. If now a thermometer capable of detecting rapid temperature vibrations is placed in the helium and moved with respect to the heater one should notice the variation of phase and amplitude characteristic of the given wave process.

In the first apparatus designed for this purpose, the rapidly varying temperature was measured by means of the method based on the detection of the alternating current on the periodically changing resistance of a bronze thermometer. Thus if the resistance of the thermometer varies according to the law $R = R_0\{1-\alpha \cos(\omega t+\varphi)\}$ (where α is small) and the voltage $E = E_0 \cos \omega t$ is applied then the current will be

$$I = \frac{E_0 \cos \omega t}{R_0\{1-\alpha \cos(\omega t+\varphi)\}} \approx \frac{E_0}{R_0} \cos \omega t \{1+\alpha \cos(\omega t+\varphi)\},$$

$$I = \frac{E_0}{R_0} \text{ (variable part)} + \frac{\alpha E_0}{2R_0} \cos \varphi,$$

i.e. a direct current arises in the circuit which depends not only on the coefficient α, which determines the amplitude of the temperature oscillations, but also on the phase difference of the oscillations of the temperature and alternating current.

A schematic drawing of the apparatus is shown in Fig. 2. The current from an audio generator of frequency $\omega/2$ is fed to the heater which is a flat mesh made of a 20-micron constantan wire, the second sound with a frequency ω is thus excited in helium II. This current simultaneously is fed to a resonance frequency doubler and therefrom into the resistor of the thermometer. The latter consists of a 0·3 mm spiral made of 30-micron phosphor-bronze wire wound in zig-zag fashion on the rim of a fiber ring.

The resonance circuit in the anode circuit of 6C5 tube selected out vibrations with a frequency ω and a condenser C permitted us to vary the frequency of the contour which in turn led to a change in the phase difference of the incident and double frequency oscillations. Control of operation of the frequency doubler was

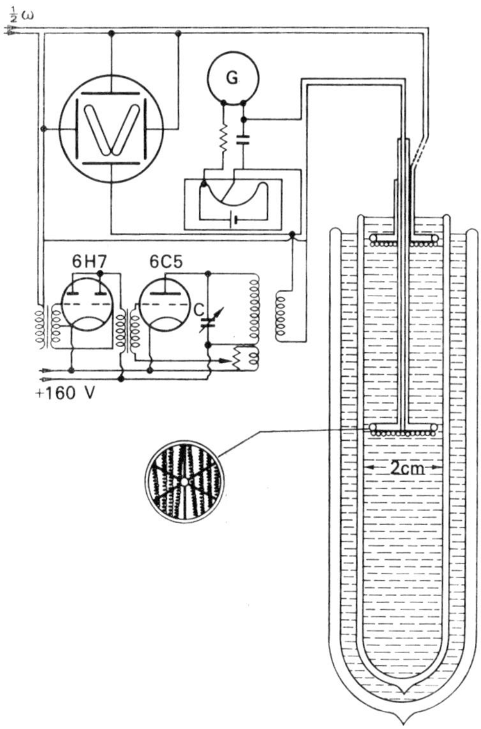

FIG. 2. First apparatus for the observation of the velocity of the second sound in helium II.

carried out by observing the Lissajous figures on a cathode-ray oscillograph.

As harmonics were present in the current supplied to the thermometer a direct current was present even when the heater was turned off. The fundamental oscillation caused periodic changes in temperature of the thermometer with a double frequency while the overtone with a frequency equal to the double one was detected by the varying resistance according to above mentioned formula. The self-detected current was thus compensated by a potentiometer and the difference in the readings of the galvanometer G obtained when the heater was turned on or off was then observed. In order to prevent undesirable vibrations of the galvanometer mirror a large resistance was connected in series with the galvanometer and both were shunted by a large capacitance as shown in Fig. 2.

By moving the thermometer with respect to the heater a periodically changing reflection of the galvanometer was really detected

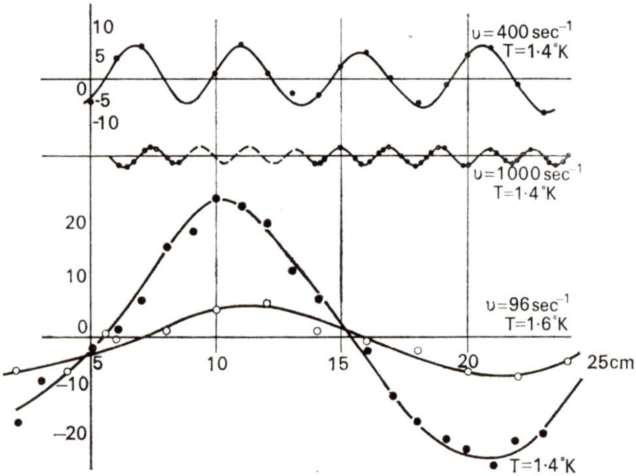

FIG. 3. Dependence of the galvanometer deflection upon the distance of the thermometer from the heater.

which confirms the existence of the second sound in helium II. It should be noted, however, that these deviations were very small and the accuracy of the measurements is insignificant. The results of some of these measurements are illustrated in Fig. 3. A short communication was previously published[13].

More accurate data were obtained by observing standing waves when the temperature vibrations were considerably amplified due to resonance. However, on determining the resonance point the rapidity of the readings was limited by the period of the galvanometer and it was also necessary to observe simultaneously the phases of the variation of the original current and the double fre-

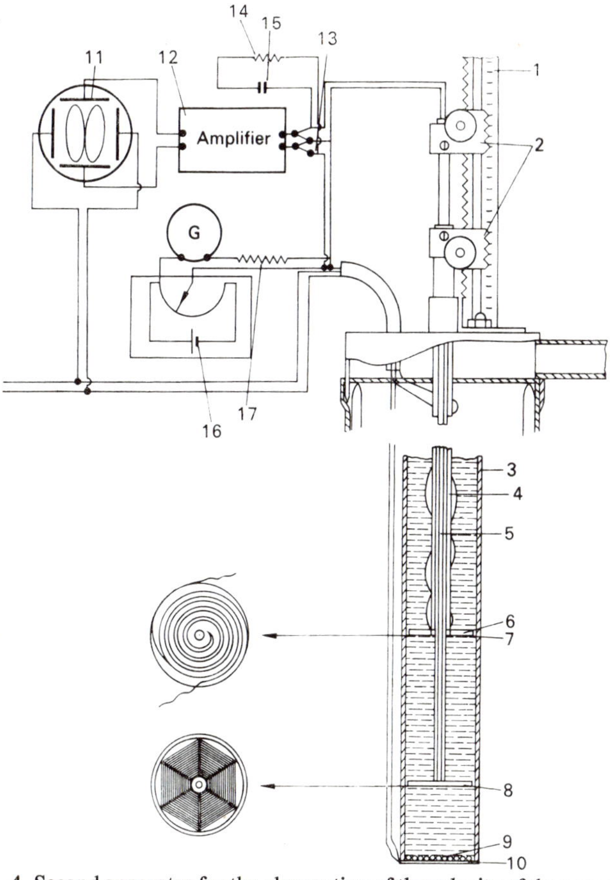

FIG. 4. Second apparatus for the observation of the velocity of the second sound in helium II.

quency current. Owing to this the main measurements were carried out on another apparatus which is schematically shown in Fig. 4.

A glass tube *3*, 2 cm in diameter and 25 cm in length, was set in a Dewar vessel containing helium II. The bottom of the tube was covered with a flat glass disc *10*. A heater *7* and resistance thermometers *8* and *9* were placed within the tube. The heater consisted of a 20-micron constantan wire wound in the form of a bifilar spiral on the plane of an ivory disc *6*. The thermometer *8* consisted of 30-micron spiral of phosphor-bronze wire also wound in the form of a bifilar spiral on the plane of an ivory ring. It was very similar to a small cobweb, being almost as thin as the latter but considerably more dense.

The disc *6* was fastened to a thin-walled steel tube *4* and could be easily moved along the glass tube. The gap between the walls of the tube and the disc was approximately 0·1 mm. The resistance thermometer *8* was fastened to a 2 mm steel tube *5* and also could be moved along the glass tube. Due to the lath *1* and verniers *2* the position of the thermometer and heater could be fixed with an accuracy up to 0·1 mm.

The thermometer *9*, a zig-zag shaped spiral of 0·3 mm diameter made of 40-micron phosphor-bronze wire was attached to the bottom.

Detection of the temperature variations in this apparatus took place in a different manner than in the first apparatus. The current from a battery *15* was led through a large resistance *14* and then across the thermometers *8* and *9*. As the resistance *14* was considerably higher than the resistance of the thermometers the current was practically constant. For periodic variations of the temperature the resistance of the thermometers and thus the voltage drop on them also varied periodically. By means of switch *13* these voltage variations either from thermometer *8* or thermometer *9* were supplied to the amplifier and then to one of the pair of deflecting plates of a cathode-ray oscillograph *11*. The voltage applied to the heater *7* was applied to the second pair of plates. The Lissajous figures were again observed.

The absence of inertia in this method is its chief advantage as the time of the experiment may thus be considerably lowered. The sensitivity is limited mainly by the noises in the tubes and self-excitation of the amplifier. As an appreciable deflection of the ray of the oscillograph requires a voltage of about 10–20 V while the background of the tubes and resistors is of the order of 1–2 μV one could raise the amplification up to about 10 million times and this was in fact carried out.

Helium II in the glass tube in the volume closed by the discs *6* and *10* formed a resonator similar to that formed by the air column in Quincke's experiments. Therefore, when the circuit is switched on and the distance between the disc and the heater is changed, very sharp resonance could be detected when an integer number of half-waves was fitted into the tube. The amplitude yields a very definite kick and the phase varies very rapidly. The maximum was so sharp that a 0·3% change in the distance between the discs leads to a twice decrease of the amplitude. At the end of tube near the glass disc and the ivory disc a loop of temperature oscillations and node of the heat flow exist which means that on the boundary with a solid body not only is $\varrho_n v_n + \varrho_s v_s = 0$ but $\varrho_n v_n = 0$ also.

As the most advantageous position for the thermometer is that of the peak of the temperature vibrations it will be seen that the phosphor-bronze thermometer *9* turned out to be just in this position.

The velocity was measured by moving the thermometer *8* and noting the nodes and the peaks as well as by registering the successive positions of the disc *6* when tuning in resonance. In the latter case the presence of the thermometer *8* within the volume did not affect in any way the results of the measurements.

An experiment was also carried out in which 6 cm of the lower part of the tube was filled up with cotton-wool. In this case a travelling wave was propagated down from the heater and absorbed by the cotton-wool, this easily being detected by moving the thermometer *8*. The amplitude of the oscillations remained constant while the phase gradually changed. The wave lengths for travelling and standing waves of the same frequency were found to coincide.

The temperature dependence of the velocity of the second sound in helium II was measured with frequencies of the alternating current of 200 and 800 cycles per second.

The frequency of the audio generator was controlled by means of a cathode-ray oscillograph and tuning-fork generator. The latter was compared with an exact clock by means of a synchronous motor and a revolution indicator. The frequency could thus be determined and maintained with an accuracy of 0·1–0·2%.

The temperature was measured according to the vapour pressure of helium. Besides, the resistance of the thermometer was determined by means of the potentiometer *16* and the temperature could then be found. The relative accuracy of the temperature measurements with the bronze thermometer was considerably higher than that obtained by observing the vapour pressure of helium; one ten-thousandth fraction of a degree could easily be detected.

For simultaneous measurement of the resistance and observation of the resonance points it was necessary that the potentiometer only slightly decreases the voltage vibrations in the bronze thermometer and this could easily be achieved by connecting a high resistance *17* in the circuit (6000 Ω as compared with 300 Ω of the thermometer).

Far from the λ-point the velocity weakly varies with temperature and the measurements were carried out by finding the resonance position of the disc *6*. Near the λ-point the velocity strongly depends on the temperature and, therefore, the resonance peaks were recorded as the temperature was slowly raised, the frequency and distance between the discs remaining constant.

In the immediate vicinity of the λ-point the resonance passes so rapidly that the temperature could be measured only by observing the deflection of the galvanometer. The potentiometer was usually compensated somewhere near 2·14–2·15°K which served as the first mark point. Then as the temperature was raised the increasing deflection of the galvanometer was measured. At the λ-point due to the sharp change in heat transfer which occurs on transition from helium II to helium I a sharp jump in the resistance took place. This jump was taken to occur at 2·19°K and served as a second mark point.

The frequency ν and length l, which contained n half-waves were the quantities directly measured in the experiments. The velocity was calculated according to the formula

$$u = 2l\nu/n.$$

Results of measurements of the velocity of the second sound by the method described above are given in Table 1 and are represented by the curve in Fig. 5 (points *1* and *2*).

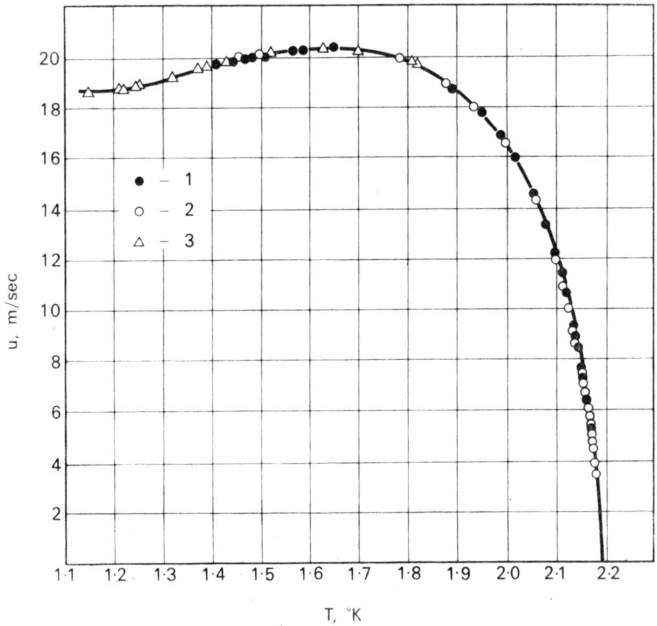

FIG. 5. Dependence of the velocity of the second sound upon temperature under the pressure of the saturated vapour.

Between 1·36 and 2·0°K the accuracy of the velocity measurements is 0·3%. At higher temperatures the velocity rapidly changes and the accuracy considerably drops due to errors in the temperature measurements.

TABLE 1

Experiment January 1, 1945		Experiment January 16, 1945	
T(°K)	u (m/sec.)	T(°K)	u (m/sec.)
1·36	19·52	1·46	20·00
1·41	19·73	1·50	20·12
1·44	19·81	1·63	20·36
1·47	19·91	1·79	19·88
1·49	19·98	1·88	18·88
1·51	20·05	1·936	17·96
1·57	20·20	2·00	16·53
1·59	20·25	2·063	14·18
1·63	20·31	2·101	12·0
1·65	20·36	2·116	10·9
1·70	20·25	2·127	10·0
1·795	19·84	2·134	9·2
1·896	18·78	2·139	8·6
1·954	17·80	2·143	8·0
1·99	16·94	2·150	7·5
2·002	16·02	2·157	7·1
2·058	14·56	2·162	6·7
2·083	13·35	2·165	6·3
2·102	12·32	2·168	6·0
2·116	11·45	2·171	5·7
2·126	10·70	2·173	5·4
2·133	10·01	2·174	5·2
2·139	9·43	2·176	5·0
2·144	8·9	2·177	4·8
2·148	8·4	2·178	4·6
2·152	8·0	2·179	4·4
2·155	7·6	2·180	4·3
2·158	7·3	2·181	4·2
2·160	7·0	2·182	3·9
2·162	6·4	2·183	3·6
2·164	6·2	2·184	3·4
2·165	5·9		

The maximum value of the velocity was determined separately with the following results: we obtained in the experiment of January 11, 1945, $T = 1{\cdot}63°K$, $u = 20{\cdot}33$ m/sec.; in the experiment of February 2, 1945, $T = 1{\cdot}63°K$, $u = 20{\cdot}31 \pm 0{\cdot}003$ m/sec.*

To measure the velocity of the second sound at lower temperatures a smaller Dewar vessel was used. A glass tube 15 cm long and 1 cm in diameter was inserted into the vessel. The ends of the tube were covered by two flat glass discs. The heater was placed at one end, the resistance thermometer at the other. Helium in the tube flowed through the slit between the edge of the tube and the glass discs.

The velocity was again determined according to the resonance peaks but this time resonance was attained by varying the frequency of the audio generator.

Results of the measurements are given in Table 2 and plotted in Fig. 5 (points *3*).

An attempt was made to detect dispersion at the temperature of the maximum velocity, i.e. at $T = 1{\cdot}63°K$, however, in the frequency range from 100 to 10,000 cycles per second it was found that the velocity of the second sound remains constant, at any rate to within an accuracy up to 0·1 %.

The obtained results permit one to determine the ratio ϱ_n/ϱ. According to Landau's theory[7] the velocity of the second sound in helium II equals

$$u = \sqrt{\frac{S^2 T \varrho_s}{C \varrho_n}},$$

where C—specific heat per unit mass, T—temperature, S—entropy of helium II per unit mass, ϱ—density of the superfluid part, ϱ_n—density of the normal part. Taking into account that $\varrho_s + \varrho_n = \varrho$ is the density of helium one obtains $\varrho_n/\varrho = 1/(1 + u^2 C/S^2 T)$.

* *(Note by Editor.)* Measurements of the velocity of the second sound below 1°K show an increase of u_2 with decreasing temperature (Atkins and Osborne, 1950; Peshkov, 1960).

If the values for C are taken from the curve plotted on the basis of the data of Keesom and Miss Keesom[14] the values of S from the curve obtained on basis of Kapitza's data[4] and the values for ϱ_n/ϱ obtained from the curve of Fig. 5 are all inserted in the latter formula one obtains the results given in Table 3 and plotted

TABLE 2

Expreiment April 6, 1945

T(°K)	u (m/sec.)
1·15	18·7
1·21	18·7
1·22	18·75
1·24	18·8
1·25	18·95
1·32	19·2
1·37	19·6
1·39	19·7
1·43	19·9
1·52	20·4
1·63	20·3
1·70	20·3
1·74	20·2
1·79	20·0
1·81	19·85
1·82	19·7

in Fig. 6. For the sake of comparison the results of Andronikashvili[12] obtained with a pile of oscillating discs in helium II are also plotted on the graph.

Figure 6 shows that the values of ϱ_n/ϱ obtained by quite different methods coincide within the accuracy of the experimental errors. The errors are about 5–6% for the calculated curve and about 4% for the data of Andronikashvili.

The data also permit one to calculate the minimum energy

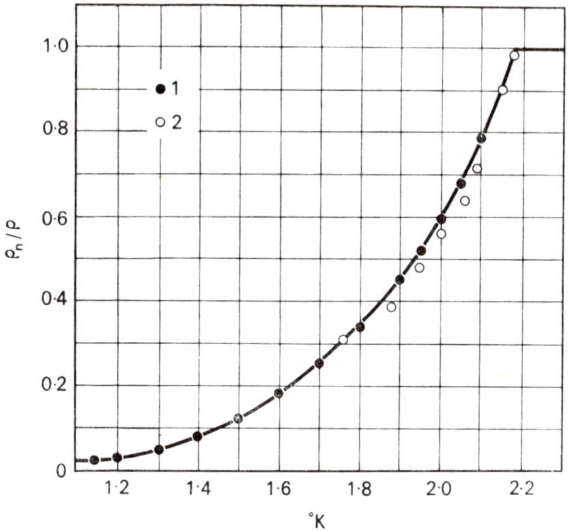

1 — values calculated by means of the formula

$$\frac{\varrho n}{\varrho} = \left(1 + \frac{u^2 C}{S \cdot T}\right)^{-1};$$

2 — values measured by Andronikashvili

FIG. 6. Dependence of ϱ_n/ϱ upon temperature.

TABLE 3

T(°K)	ϱ_n/ϱ	Δ/k	μ/m_{He}
1·3	0·047	9·42	13·1
1·4	0·076	9·35	12·1
1·5	0·119	9·32	11·6
1·6	0·183	9·30	11·3
1·7	0·252	9·13	10·3
1·8	0·336	9·09	9·8
1·9	0·450	9·00	9·4
1·95	0·518		
2·0	0·593	8·99	9·3
2·05	0·680		
2·1	0·785	8·80	8·9
2·15	0·901		

Δ required for the production of rotons, and the effective mass μ of rotons.

According to Landau[7] the entropy of the roton part

$$S_r = \frac{1}{\varrho}\left(\frac{\mu k}{2\pi\hbar^2}\right)^{1/2} T^{2/1}\, \Delta e^{-\Delta/kT}\left(1-\frac{5kT}{2\Delta}\right)$$

and

$$\frac{\varrho_n}{\varrho} = \frac{\mu}{\varrho}\left(\frac{\mu kT}{2\pi\hbar^2}\right)^{3/2} e^{-\Delta/kT}.$$

Combining these equations we obtain

$$S_r = \frac{k}{\varrho}\left(\frac{\varrho_n kT\varrho}{\varrho 2\pi\hbar^2}\right)^{3/2} e^{-\Delta/kT}\left(\frac{\Delta}{kT}+\frac{5}{2}\right), \tag{1}$$

$$\frac{\mu}{m_{\mathrm{He}}} = \frac{k}{m_{\mathrm{He}}}\,\frac{\varrho_n}{\varrho}\,\frac{\left(\frac{\Delta}{kT}+\frac{5}{2}\right)}{S_r}, \tag{2}$$

where m_{He} is the mass of the helium atom. The roton part of the entropy can be found as the difference between the total entropy and its phonon part. The latter equals, according to Landau's theory

$$S_{\mathrm{ph}} = \frac{\pi^2 k^4}{90\hbar^3\varrho}\,\frac{\partial}{\partial T}\left(\frac{T^4}{c^3}\right) = 1.8\cdot 10^{-3}\, T^3 \text{ cal./degree,}$$

where c is velocity of sound

$$S_r = S - S_{\mathrm{ph}}.$$

If S is taken from Kapitza's data and ϱ_n/ϱ from Table 3, then one obtains from equation (1) for Δ/k the results set out in Table 3. The values of Δ/k are expressed in degrees. The accuracy of determination of Δ/k is approximately 4%.

On the basis of the results obtained here it may be considered that at low temperatures when the density of the roton gas is small Δ/k is constant and equals 9·3°. The decrease of Δ/k which begins at 1·7°K should be ascribed to the increase of the interaction between the rotons as their density is increased and hence to the

invalidity of formula (1) which was derived on the assumption of a perfect roton gas.

Putting $\Delta/k = 9{\cdot}3°$ and using the same values for S and ϱ_n/ϱ as were used in the preceding computation, we obtain from formula (2) the values for μ/m_{He} which are given in the last column of Table 3.

The effective roton mass is expressed in units of the mass of the helium atom and was determined with an accuracy of about 10%.

As in the preceding case at temperatures near the λ-point the formulae are invalid and the results should sooner indicate the degree of deviation of the roton gas from an ideal one rather than the variation of the mass with temperature.

Experiments to determine the velocity at high pressures and also to detect in the second sound a phenomenon similar to the critical velocities observed by Kapitza[4] in his experiments are now in progress.

CONCLUSIONS

The present experiments show that due to the peculiar mode of the heat transfer in helium II, thermal vibrations are propagated not in the form of ordinary rapidly damped heat waves but as undamped vibrations which Landau called the second sound.

The temperature dependence of the velocity of the second sound fully confirms the phenomenological part of the theory of superfluidity and in particular the hydrodynamics of helium II, which predicts the possibility of two simultaneous motions—the normal and superfluid ones.

The microscopic theory agrees with the experiments less satisfactorily. Calculations on the basis of the present data of the energy gap between the potential and vortex motions of helium II, or, what is the same, the minimum energy required for the production of a roton Δ and also calculation of the effective mass of the roton μ at low temperatures (from 1·3° to 1·6°K) yield values which though varying less than the errors of their determination still manifest a marked temperature dependence.

No special arguments exist which favour the assumption that the roton mass should remain constant for any temperature, all the more it is clear that near the λ-point μ should charge due to the high density of the gas of heat quanta.

Only after experiments at lower temperatures will be carried out will it be possible to determine whether μ and Δ are constant at temperatures below 1·5°K or whether the microscopic theory of helium II should be modified in some manner.

After the completion of the present work two papers by Tisza(15,16) were received in which a theory of helium II was developed on the basis of the assumption of the possibility of dividing the helium atoms into excited and unexcited ones and the possibility of free movement of one kind of atoms with respect to the others. These assumptions seem to be too simplified and although Tisza explains some of the facts qualitatively satisfactorily, such as, for example, the existence of temperature waves, i.e. second sound, quantitatively the theory yields incorrect results.

Superconductivity for electrons in metals is the analogy of superfluidity. The superconducting electrons also possess zero entropy while the excited ones contain the full heat energy. One might, therefore, expect to find in superconductors some counterpart for the second sound, at any rate in the region in which the electron thermal capacity is greater than that of the lattice. However, as in the second sound the part with zero entropy oscillates in a direction opposite to that of the normal part, the decay should be considerably due to friction between the normal electrons and the lattice. Thus in superconductors one should expect only a deviation from the law of propagation of ordinary heat waves, the damping being slightly less.

A perfect crystal is another type of body in which might exist the second sound. There, just as in helium II, the thermal quanta should move freely and transfer momentum to each other, in other words, a gas of thermal quanta, capable of performing vibrations similar to those of sound, should exist in a perfect crystal. However, a perfect crystal cannot be prepared, there are unavoidable inhomogeneities and irregularities which scatter the thermal quanta and as a result of which the process of heat transfer will approach the usual diffusion process.

However, at low temperatures one should expect the damping of the thermal waves in crystals being less than it ought to be, according to the laws of diffuse propagation of heat.

In conclusion I wish to express to Prof. P. L. Kapitza my sincere gratitude for suggesting to me this problem and for his constant interest in this investigation and to Prof. L. D. Landau, discussions with whom assisted me in theoretical interpretation of the experimental results. I should also like to thank E. L. Andronikashvili who read the manuscript of this paper and made some valuable suggestions.

REFERENCES

1. W. H. KEESOM and MISS KEESOM, *Physica*, **3**, 359 (1936).
2. P. KAPITZA, *Nature*, **141**, 74 (1938).
3. P. KAPITZA, *Journ. of Phys.*, **4**, 181 (1940).
4. P. KAPITZA, *Journ. of Phys.*, **5**, 59 (1941).
5. F. LONDON, *Nature*, **141**, 643 (1938).
6. L. TISZA, *Nature*, **141**, 913 (1938).
7. L. LANDAU, *Journ. of Phys.*, **5**, 71 (1941).
8. L. LANDAU, *Journ. of Phys.*, **8**, 1 (1944).
9. J. C. FINDLAY, A. PITT, H. GRAYSON SMITH and J. O. WILHELM, *Phys. Rev.*, **54**, 506 (1938).
10. E. LIFSHITZ, *Journ. of Phys.*, **8**, 110 (1944).
11. W. H. KEESOM and G. E. MACWOOD, *Physica*, **5**, 737 (1938).
12. E. ANDRONIKASHVILI, *Journ. of Phys.*, **10**, 201 (1946).
13. V. PESHKOV, *Journ. of Phys.*, **8**, 381 (1944).
14. W. H. KEESOM and MISS KEESOM, *Physica*, **2**, 557 (1935).
15. L. TISZA, *Journ. de Physique*, **1**, 164 (1940).
16. L. TISZA, *Journ. de Physique*, **1**, 350 (1940).

Theoretical Papers

T1. THE THEORY OF SUPERFLUIDITY OF HELIUM II*

L. D. LANDAU

Summary. The quantisation of an arbitrary system of interacting particles (a liquid) is performed by means of introducing the operators of the density and of the velocity of the liquid: the commutation rules between these operators are determined (§ 1). From the results of this quantisation the general character of the distribution of the energy levels in the spectrum of a quantum liquid is determined (§ 2). The temperature dependence of the heat capacity of helium II is investigated (§ 3). It is shown that at absolute zero a quantum liquid can possess the property of superfluidity (§ 4). At nonzero temperatures it is found that two motions—a superfluid and a normal—can simultaneously exist in helium II. This can be described by means of the conception of the superfluid and normal parts of the liquid; the λ-point in helium II is connected with the disappearance of the "superfluid" part of the liquid (§ 5). The experiments made to measure the heat conductivity and viscosity of helium II are interpreted; the thermomechanical effects in helium II are considered (§ 6). A system of hydrodynamic equations is advanced describing the macroscopic motion of helium II (§ 7). By means of these equations the propagation of sound is investigated and it is shown that two velocities of sound must exist in helium II (§ 8).

Liquid helium is known to possess a number of peculiar properties at temperatures lower than the λ-point. Of these properties the most important one is superfluidity discovered by P. L. Kapitza[1]—the lack of viscosity during the flow of helium through a thin capillary or slit.

All these properties, including the fact that helium exists as a liquid right down to absolute zero, obviously cannot be explained by the classical theory and are connected with quantum phenomena.

* *J. Phys. (U.S.S.R.)* **5**, 71, 1941.

L. Tisza[2] suggested that helium II should be considered as a degenerate ideal Bose gas. He suggested that the atoms found in the normal state (a state of zero energy) move through the liquid without friction. This point of view, however, cannot be considered as satisfactory. Apart from the fact that liquid helium has nothing to do with an ideal gas, atoms in the normal state would not behave as "superfluid". On the contrary, nothing could prevent atoms in a normal state from colliding with excited atoms, i.e., when moving through the liquid they would experience a friction and there would be no superfluidity at all. In this way the explanation advanced by Tisza not only has no foundation in his suggestions but is in direct contradiction to them.

1. THE QUANTISATION OF THE MOTION OF LIQUIDS

An arbitrary system of interacting particles (a liquid) can be described in classical theory by means of the density ϱ and the flow of mass $\boldsymbol{j}$, which are determined in the following manner. Let $\boldsymbol{R}$ be the radius-vector of an arbitrary point in space and $\boldsymbol{r}_a$—the radiusvector of a particle with a mass m_a. Then ϱ is determined as

$$\varrho = \sum_a m_\alpha \delta(\boldsymbol{r}_\alpha - \boldsymbol{R}), \tag{1.1}$$

δ being the three-dimensional δ-function and the summation is extended over all particles in the system. The volume-integral $\int \varrho dV$ gives the total mass of the system. Similarly the density $\boldsymbol{j}$ of the flow of the mass is determined as

$$\boldsymbol{j} = \sum_a m_\alpha \boldsymbol{v}_\alpha \delta(\boldsymbol{r}_\alpha - \boldsymbol{R}) = \sum_a \boldsymbol{p}_\alpha \delta(\boldsymbol{r}_\alpha - \boldsymbol{R})$$

($\boldsymbol{v}_\alpha$, $\boldsymbol{p}_\alpha$ are the velocity and the momentum of the particle m_α).

It must be emphasised that in such a description of a liquid there is no averaging in that sense in which it is done in statistics. This description proceeds from the microscopic picture as all the particles possess (at a given moment) definite coordinates $\boldsymbol{r}_\alpha$ and velocities $\boldsymbol{v}_\alpha$.

When passing over to the quantum theory ϱ and $\boldsymbol{j}$ must be regarded as certain operators the form of which must be determined. For the sake of simplicity suppose that the system consists of one particle only. Then the classical density is $\varrho = m\delta(\boldsymbol{r}-\boldsymbol{R})$. The operator ϱ must be determined in such a way that its mathematical expectation $\int\psi(\boldsymbol{r})\varrho\psi(\boldsymbol{r})dV$ [$\psi(\boldsymbol{r})$ being the wave function of the particle] equals the density of the mass at the point $\boldsymbol{R}$, i.e., $m|\psi(\boldsymbol{R})|^2$. From this it follows that the operator ϱ must have the same form $\varrho = m\delta(\boldsymbol{r}-\boldsymbol{R})$ and in the case of an arbitrary system of particles correspondingly the form (1.1).

The classical density of the flow for one particle is $\boldsymbol{j} = \boldsymbol{p}\delta(\boldsymbol{r}-\boldsymbol{R})$. It is easy to see that the corresponding operator is

$$\boldsymbol{j} = \tfrac{1}{2}[\boldsymbol{p}\delta(\boldsymbol{r}-\boldsymbol{R})+\delta(\boldsymbol{r}-\boldsymbol{R})\boldsymbol{p}],$$

where $\boldsymbol{p}$ is the usual operator of momentum:

$$\boldsymbol{p} = \frac{\hbar}{i}\nabla$$

(∇ denotes the differentiation with respect to $\boldsymbol{r}$). Actually the mathematical expectation of $\boldsymbol{j}$ is

$$\int\psi(\boldsymbol{r})^*\boldsymbol{j}\psi(\boldsymbol{r})dV = \frac{\hbar}{2i}\int\psi^*\nabla\delta(\boldsymbol{r}-\boldsymbol{R})\psi dV+\frac{\hbar}{2i}\int\psi^*\delta(\boldsymbol{r}-\boldsymbol{R})\nabla\psi dV$$

or, integrating the first term by parts:

$$\int\psi^*\boldsymbol{j}\psi\boldsymbol{j}dV = -\frac{\hbar}{2i}\int\psi\delta(\boldsymbol{r}-\boldsymbol{R})\nabla\psi^* dV+\frac{\hbar}{2i}\int\psi^*\delta(\boldsymbol{r}-\boldsymbol{R})\nabla\psi dV$$

$$= \frac{\hbar}{2i}\{\psi^*(\boldsymbol{R})\nabla\psi(\boldsymbol{R})-\psi(\boldsymbol{R})\nabla\psi^*(\boldsymbol{R})\},$$

i.e., exactly what it ought to be. For an arbitrary system we have, similarly

$$\boldsymbol{j} = \tfrac{1}{2}\sum[\boldsymbol{p}_\alpha\delta(\boldsymbol{r}_\alpha-\boldsymbol{R})+\delta(\boldsymbol{r}_\alpha-\boldsymbol{R})\boldsymbol{p}_\alpha], \tag{1.2}$$

$$\boldsymbol{p}_\alpha = \frac{\hbar}{i}\nabla_\alpha.$$

We now determine the commutation rules. For the density ϱ we obviously have

$$\varrho_1\varrho_2 - \varrho_2\varrho_1 = 0 \tag{1.3}$$

[ϱ_1, ϱ_2 denote $\varrho(\boldsymbol{R}_1)$, $\varrho(\boldsymbol{R}_2)$, respectively].

For the sake of brevity let us consider only one term from each of the sums (1.1) and (1.2) when determining the commutation rules, as the operators corresponding to different particles commute with each other. To determine the commutation of ϱ with $\boldsymbol{j}$ we write

$$\boldsymbol{j}_1\varrho_2 - \varrho_2\boldsymbol{j}_1 = \frac{m\hbar}{2i}\{[\nabla\delta(\boldsymbol{r}-\boldsymbol{R}_1)+\delta(\boldsymbol{r}-\boldsymbol{R}_1)\nabla]\delta(\boldsymbol{r}-\boldsymbol{R}_2)$$
$$-\delta(\boldsymbol{r}-\boldsymbol{R}_2)[\nabla\delta(\boldsymbol{r}-\boldsymbol{R}_1)+\delta(\boldsymbol{r}-\boldsymbol{R}_1)\nabla]\}.$$

To simplify the expression on the right-hand side we note that the operators of the form

$$\delta(\boldsymbol{r}-\boldsymbol{R}_1)\nabla\delta(\boldsymbol{r}-\boldsymbol{R}_2)$$

can be transformed in the following way:

$$\delta(\boldsymbol{r}-\boldsymbol{R}_1)\nabla\delta(\boldsymbol{r}-\boldsymbol{R}_2) = \delta(\boldsymbol{r}-\boldsymbol{R}_1)(\nabla\delta(\boldsymbol{r}-\boldsymbol{R}_2))+\delta(\boldsymbol{r}-\boldsymbol{R}_1)\delta(\boldsymbol{r}-\boldsymbol{R}_2)\nabla,$$

where in the first term $(\nabla\delta(\boldsymbol{r}-\boldsymbol{R}_2))$ denotes simply a gradient of the δ-function, i.e., ∇ is no longer an operator. Owing to the presence of the factor $\delta(\boldsymbol{r}-\boldsymbol{R}_1)$ in this term one can write the $(\nabla\delta(\boldsymbol{R}_1-\boldsymbol{R}_2))$ instead of $(\nabla\delta(\boldsymbol{r}-\boldsymbol{R}_2))$. In this way

$$\delta(\boldsymbol{r}-\boldsymbol{R}_1)\nabla\delta(\boldsymbol{r}-\boldsymbol{R}_2) = \delta(\boldsymbol{r}-\boldsymbol{R}_1)(\nabla\delta(\boldsymbol{R}_1-\boldsymbol{R}_2))+\delta(\boldsymbol{r}-\boldsymbol{R}_1)\delta(\boldsymbol{r}-\boldsymbol{R}_2)\nabla.$$

Similarly

$$\nabla\delta(\boldsymbol{r}-\boldsymbol{R}_1)\delta(\boldsymbol{r}-\boldsymbol{R}_2) = \delta(\boldsymbol{r}-\boldsymbol{R}_2)\nabla\delta(\boldsymbol{r}-\boldsymbol{R}_1)+\delta(\boldsymbol{r}-\boldsymbol{R}_1)(\nabla\delta(\boldsymbol{R}_1-\boldsymbol{R}_2)).$$

The result is

$$\boldsymbol{j}_1\varrho_2 - \varrho_2\boldsymbol{j}_1 = \frac{\hbar}{i}m\delta(\boldsymbol{r}-\boldsymbol{R}_1)\nabla\delta(\boldsymbol{R}_1-\boldsymbol{R}_2)$$

or for an arbitrary system

$$\boldsymbol{j}_1\varrho_2-\varrho_2\boldsymbol{j}_1 = \frac{\hbar}{i}\,\varrho_1\nabla\delta(\boldsymbol{R}_1-\boldsymbol{R}_2). \tag{1.4}$$

(It makes no difference whether we write ϱ_1 or ϱ_2 on the right-hand side in view of the presence of the δ-function of $\boldsymbol{R}_1-\boldsymbol{R}_2$.)

In a similar way the commutation rules between the components of the vector $\boldsymbol{j}$ with each other can be obtained. The calculation in this case is longer and we will not enter into it here.

We introduce the operator $\boldsymbol{v}$ of the velocity of the liquid according to

$$\boldsymbol{j} = \tfrac{1}{2}(\varrho\boldsymbol{v}+\boldsymbol{v}\varrho) \tag{1.5}$$

$$\boldsymbol{v} = \frac{1}{2}\left(\frac{1}{\varrho}\boldsymbol{j}+\boldsymbol{j}\frac{1}{\varrho}\right). \tag{1.6}$$

It will be more convenient to use the operator $\boldsymbol{v}$ instead of the operator of the flow $\boldsymbol{j}$.

For the commutation rule of ϱ with $\boldsymbol{v}$ we have

$$\begin{aligned}\boldsymbol{v}_1\varrho_2-\varrho_2\boldsymbol{v}_1 &= \frac{1}{2}\left(\frac{1}{\varrho_1}\boldsymbol{j}_1+\boldsymbol{j}_1\frac{1}{\varrho_1}\right)\varrho_2-\frac{1}{2}\varrho_2\left(\frac{1}{\varrho_1}\boldsymbol{v}_1+\boldsymbol{v}_1\frac{1}{\varrho_1}\right)\\ &= \frac{1}{2\varrho_1}(\boldsymbol{j}_1\varrho_2-\varrho_2\boldsymbol{j}_1)+\frac{1}{2}(\boldsymbol{j}_1\varrho_2-\varrho_2\boldsymbol{j}_1)\frac{1}{\varrho_1}\end{aligned}$$

or, on inserting (1.4):

$$\boldsymbol{v}_1\varrho_2-\varrho_2\boldsymbol{v}_1 = \frac{\hbar}{i}\nabla\delta(\boldsymbol{R}_1-\boldsymbol{R}_2). \tag{1.7}$$

The commutation rules for the components of $\boldsymbol{v}$ are found to be*

$$v_{1i}v_{2k}-v_{2k}v_{1i} = \frac{\hbar}{i}\,\delta(\boldsymbol{R}_1-\boldsymbol{R}_2)\frac{1}{\varrho_1}(\mathrm{curl}\,\boldsymbol{v})_{ik}, \tag{1.8}$$

* *(Note by Editor)*. The correctness of (1, 8) was disputed by F. London [*Rev. Mod. Phys.* **17**, 310 (1945)]. The proof of this formula was given by Ziman (1953).

where $(\text{curl } \boldsymbol{v})_{ik}$ denotes the difference

$$\frac{\partial v_k}{\partial x_i} - \frac{\partial v_i}{\partial x_k}.$$

Further on we shall also need the commutation rule between ϱ and curl $\boldsymbol{v}$. By applying the operation curl (with a differentiation with respect to coordinates $\boldsymbol{R}_1$) to both sides of the equation (1.7) we get

$$\text{curl } \boldsymbol{v}_1 \cdot \varrho_2 - \varrho_2 \cdot \text{curl } \boldsymbol{v}_1 = 0 \tag{1.9}$$

It is easy to see that by applying the formulae obtained to the macroscopic movement of the liquid we get, as required, the usual hydrodynamic equations written in an operational form. The energy of a unit volume of a classical liquid considered macroscopically is

$$\frac{\varrho v^2}{2} + \varrho\varepsilon(\varrho),$$

where $\varepsilon(\varrho)$ is the internal energy of a unit mass of the liquid. It is supposed that the energy ε depends only on the density ϱ of the liquid; this corresponds to the macroscopic character of the argument and is connected with a statistical averaging. For a microscopic investigation this supposition is, of course, invalid.

The corresponding quantum operators*

$$\frac{\boldsymbol{v}\varrho\boldsymbol{v}}{2} + \varrho\varepsilon(\varrho).$$

The Hamiltonian H of the liquid is an integral over the volume

$$H = \int \left\{ \frac{\boldsymbol{v}\varrho\boldsymbol{v}}{2} + \varrho\varepsilon(\varrho) \right\} dV. \tag{1.10}$$

For the derivative of the density ϱ with respect to time one has

$$\dot{\varrho} = \frac{i}{\hbar}(H\varrho - \varrho H).$$

* The operator $\boldsymbol{v}\varrho\boldsymbol{v}/2$ can also be written in the form

$$[\varrho \boldsymbol{v}^2 + \boldsymbol{v}^2 \varrho]/4$$

We shall denote temporarily the coordinates of the point at which ϱ is taken by the index 1, and the coordinates of the variable point in the region of integration in (1.10) by the index 2. Then

$$\dot{\varrho}_1 = \frac{i}{\hbar}\int\left\{\frac{1}{2}[\boldsymbol{v}_2\varrho_2\boldsymbol{v}_2\varrho_1-\varrho_1\boldsymbol{v}_2\varrho_2\boldsymbol{v}_2]+[\varrho_2\varepsilon(\varrho_2)\varrho_1-\varrho_1\varrho_2\varepsilon(\varrho_2)]\right\}dV_2$$

In view of (1.3) the second term under the sign of the integration vanishes, and the first can be written as

$$\tfrac{1}{2}[\boldsymbol{v}_2\varrho_2(\boldsymbol{v}_2\varrho_1-\varrho_1\boldsymbol{v}_2)+(\boldsymbol{v}_2\varrho_1-\varrho_1\boldsymbol{v}_2)\varrho_2\boldsymbol{v}_2]$$

or, by introducing (1.7):

$$\frac{\hbar}{2i}\nabla\delta(\boldsymbol{R}_2-\boldsymbol{R}_1)(\boldsymbol{v}_2\varrho_2+\varrho_2\boldsymbol{v}_2) = \frac{\hbar}{i}\nabla\delta(\boldsymbol{R}_2-\boldsymbol{R}_1)\boldsymbol{j}_2,$$

In this way

$$\dot{\varrho}_1 = \int\nabla\delta(\boldsymbol{R}_2-\boldsymbol{R}_1)\boldsymbol{j}_2 dV_2 = -\int\delta(\boldsymbol{R}_2-\boldsymbol{R}_1)\,\mathrm{div}\,\boldsymbol{j}_2 dV_2 = -\mathrm{div}\,\boldsymbol{j}_1,$$

i.e., we come to the continuity equation in operational form:

$$\frac{\partial\varrho}{\partial t}+\mathrm{div}\,\frac{\varrho\boldsymbol{v}+\boldsymbol{v}\varrho}{2} = 0. \qquad (1.11)$$

In a similar way the derivative:

$$\dot{\boldsymbol{v}} = \frac{i}{\hbar}(H\boldsymbol{v}-\boldsymbol{v}H),$$

can be calculated, which brings us to the equation

$$\frac{\partial v_i}{\partial t}+\frac{1}{2}\left(v_k\frac{\partial v_i}{\partial x_k}+\frac{\partial v_i}{\partial x_k}v_k\right) = -\frac{1}{\varrho}\frac{\partial}{\partial x_i}\frac{\partial\varepsilon}{\partial\varrho}, \qquad (1.12)$$

i.e., Euler's equation in an operational form ($d\varepsilon/d\varrho$ is the pressure p of the liquid).

It must be again emphasised that the equations (1.11), (1.12) are less general than the commutation rules (1.3)–(1.9), which are also valid for an exact, microscopical investigation of the liquid.

2. THE ENERGY SPECTRUM OF A QUANTUM LIQUID

In the classical hydrodynamics of ideal liquids it is shown that if, at a certain moment of time, the motion is potential (curl $\boldsymbol{v} = 0$) in the whole volume of the liquid, it will be potential for all other moments of time (Lagrange's theorem). It appears that this classical theorem finds its analogy in quantum hydrodynamics.

According to the commutation rules (1.9), curl $\boldsymbol{v}$ always commutes with the density ϱ. The components of curl $\boldsymbol{v}$, however, do not commute, generally speaking, either with each other or with the components of velocity $\boldsymbol{v}$ [when the operation curl is applied to the equation (1.8) the right-hand side does not vanish]. Therefore, curl $\boldsymbol{v}$ does not, generally speaking, commute with the Hamiltonian, i.e., is not conserved.

An exception is the case when over the whole volume of the liquid curl $\boldsymbol{v} = 0$. In this case we have zero in the right-hand side of (1.8) and curl $\boldsymbol{v}$ commutes with ϱ and $\boldsymbol{v}$ and, therefore, also with the Hamiltonian.†

In this way curl $\boldsymbol{v}$ is conserved if it is zero. In other words, a quantum liquid always possesses stationary states in which curl $\boldsymbol{v}$ equals zero over the whole volume of the liquid. Such a state might be called, by analogy to classical hydrodynamics, a state of potential motion of the liquid.

Concerning these results an analogy can be made with the angular momentum $\boldsymbol{M}$ in quantum mechanics. The commutation of two components of $\boldsymbol{M}$ with each other leads to the third component of $\boldsymbol{M}$, with the result that all the components of $\boldsymbol{M}$ commute with each

† Not only with the Hamiltonian (1.10), but also with all other functions containing $\boldsymbol{v}$, ϱ and their derivates of any order with respect to the coordinates.

other if they are all equal to zero. It is also known that there exist no states with an infinitely small angular momentum, its first nonzero eigenvalues are of the order of $\hbar$. This is a consequence of the fact that the commutation rules are inhomogeneous—their left-hand sides are quadratic in $\boldsymbol{M}$ and the right-hand sides are linear.

A similar statement can be advanced concerning curl $\boldsymbol{v}$ in quantum hydrodynamics. Namely, no states can exist in which curl $\boldsymbol{v}$ would be non-zero, but arbitrarily small over the whole volume of the liquid. In other words, between the states of the potential (curl $\boldsymbol{v} = 0$) and vortex (curl $\boldsymbol{v} \neq 0$) motions of a quantum liquid there is no continuous transition.

From this the principal features of the energy spectrum of a liquid directly follow. The presence of a gap between the states of the potential and vortex motions means that between the lowest energy levels of vortex and potential motions a certain finite energy interval must exist. As to the question which of these two levels lies lower, apparently both cases are logically possible. It will be shown below that we get the phenomenon of superfluidity if we suppose that the normal level of the potential motions lies lower than the normal level of vortex motions. Hence we must suppose that this very case exists in liquid helium. It must be remarked, however, that, as only one quantum liquid exists, liquid helium, the question as to whether such a distribution of the levels and hence the property of superfluidity is a general property of a quantum liquid cannot be solved experimentally.

This brings us to the following picture of the distribution of the levels in the energy spectrum of liquid helium (it must be emphasised that we do not here refer to the levels for single helium atoms but to the levels corresponding to the states of the whole liquid). This spectrum is made up of two superimposed continuous spectra. One of them corresponds to the potential motions and the other—to vortex motions. The lowest level of the vortex spectrum is situated above the lowest level of the potential spectrum, this latter level being the normal unexcited state of a liquid; the energy interval between these two levels we denote by Δ.

The value of the energy gap Δ cannot be calculated exactly. Its order of magnitude is

$$\Delta \sim \frac{\hbar^2 \varrho^{2/3}}{m^{5/3}} \tag{2.1}$$

(m being the mass of the helium atom and ϱ—the density of the liquid). This is the only quantity of the dimension of energy which can be built up from m, ϱ and $\hbar$. This gives numerically $\Delta/k \sim 1°$, i.e., Δ, as was expected, is of the order of kT_λ, T_λ being the temperature of the λ-point of helium [cf. (3.8)].

Consider an excited level which is situated not too high above the beginning of the spectrum (vortex of potential one).

Every weakly excited state can be considered as an aggregate of a number of single "elementary excitations". As far as the excited levels of the potential spectrum are concerned, the potential internal motions of the liquid are longitudinal waves, i.e., these motions are sound waves. Therefore, the corresponding elementary excitations are simply sound quanta, i.e., phonons. The energy of the phonons is known to be a linear function of their momentum p:

$$\varepsilon = cp, \tag{2.2}$$

c being the velocity of sound. Thus, at the beginning of the potential spectrum the energy is proportional to the first power of the momentum.

An "elementary excitation" of the vortex spectrum might be called a "roton".* Those special reasons which stipulate a linear dependence of ε on p for phonons do not exist for rotons. For small momenta $\boldsymbol{p}$ the energy of the roton can be simply expanded in powers of $\boldsymbol{p}$; in view of the isotropy of the liquid the expansion of the scalar ε in powers of the vector $\boldsymbol{p}$ only contains terms with

* This name was suggested by I. E. Tamm.

even powers, so one may write

$$\varepsilon = \Delta + \frac{p^2}{2\mu}, \tag{2.3}$$

where μ is an "effective mass" of the roton [in (2.2) and (2.3) the energy is measured from the normal state].*†

If the number of phonons and rotons (per unit volume of the liquid) is not large, their aggregate can be regarded as a mixture of two ideal gases—a phonon gas and a roton gas. It is known that the phonon gas obeys Bose statistics. As to the rotons, they too probably obey Bose statistics. It must, however, be remarked that inasmuch as the energy of a roton always contains a quantity Δ large compared with kT (at low temperatures only when the aggregate of rotons can be treated as a gas) the difference between the Bose and Fermi statistics is not essential and one can use Boltzmann's distribution for the rotons.

3. THE HEAT CAPACITY OF HELIUM II

From the properties of the energy spectrum one can come to certain conclusions concerning the temperature dependence of the heat capacity of helium II.

At sufficiently low temperatures, when the abnormal behaviour of the heat capacity near the λ-point is no longer essential, the heat capacity may be considered as consisting of two parts—a phonon and a roton part. The first (C_{ph}) is no other than the Debye heat capacity proportional to the cube of the temperature. This part of the heat capacity can be calculated exactly‡ as the Debye

* See Landau's later note (paper T3 in this volume), formula (2).

† In a recent paper A. Bijl[3] investigated the properties of the energy spectrum of a liquid and came to the conclusion that there must be an energy gap between the normal and all excited states. This result does not seem to be plausible as it would mean, in particular, the impossibility of the propagation of sound waves with small frequencies in liquids.

‡ This calculation had already been made in 1940 by A. Migdal whom I wish to thank for informing me of his results.

temperature can be calculated from the data obtained[4] for the compressibility of helium from which, in its turn, the velocity of sound in helium can be calculated (it turns out to be 250 m/sec). According to the known formula we get, for the heat capacity per 1 g of helium

$$C_{\mathrm{ph}} = 4{\cdot}4\times 10^{-3} T^3 \ \mathrm{cal/^\circ Kg}. \tag{3.1}$$

The roton part of the heat capacity (C_r) has, owing to the presence of the energy gap Δ between the normal and vortex levels, an exponential temperature dependence. The coefficient before the exponential factor can be expressed through Δ and the effective mass μ of the roton. For the temperatures considered one can, as has already been pointed out, apply Boltzmann's distribution to the roton gas. The free energy of the gas with an undetermined number of particles (the number of particles in the roton gas is itself a temperature function determined from the condition of the minimum of the free energy) is

$$F_{\mathrm{r}} = -kTV \int e^{-\varepsilon/kT} \frac{d\tau_p}{(2\pi\hbar)^3}$$

$$(d\tau_p = dp_x dp_y dp_z)$$

(V is the volume). Below we shall refer the free energy to 1 g of helium corresponding to which we shall put $V = 1/\varrho$ (ϱ being the helium density), so that

$$F_{\mathrm{r}} = -\frac{kT}{\varrho(2\pi\hbar)^3} \int e^{-\varepsilon/kT} d\tau_p. \tag{3.2}$$

The number of rotons is

$$N_{\mathrm{r}} = \frac{1}{\varrho(2\pi\hbar)^3} \int e^{-\varepsilon/kT} d\tau_p. \tag{3.3}$$

Putting $\varepsilon = \Delta + p^2/2\mu$ we get

$$F_{\mathrm{r}} = -\frac{kT}{\varrho}\left(\frac{\mu kT}{2\pi\hbar^2}\right)^{3/2} e^{-\Delta/kT}; \tag{3.4}$$

$$N_{\mathrm{r}} = \frac{1}{\varrho}\left(\frac{\mu kT}{2\pi\hbar^2}\right)^{3/2} e^{-\Delta/kT}. \tag{3.5}$$

Thus the entropy is

$$S_{\mathrm{r}} = -\frac{\partial F_{\mathrm{r}}}{\partial T} = \frac{1}{\varrho}\left(\frac{\mu k}{2\pi\hbar^2}\right)^{3/2} T^{1/2}\Delta e^{-\Delta/kT}\left(1+\frac{5kT}{2\Delta}\right), \qquad (3.6)$$

and the heat capacity is

$$C_{\mathrm{r}} = T\frac{\partial S_{\mathrm{r}}}{\partial T} = \frac{1}{\varrho}\left(\frac{\mu}{2\pi\hbar^2}\right)^{3/2} \frac{k^{1/2}\Delta^2}{T^{1/2}} e^{-\Delta/kT}\left[1+3\frac{kT}{\Delta}+\frac{15}{4}\left(\frac{kT}{\Delta}\right)^2\right]. \qquad (3.7)$$

If we use the data obtained by Keesom and Miss Keesom[5] for the heat capacity in the interval 1·2–1·5°K from the formulae derived we get for μ and Δ the values

$$\frac{\Delta}{k} = 8\text{–}9°\mathrm{K}, \quad \mu = 7\text{–}8 m_{\mathrm{He}} \qquad (3.8)$$

(m_{He} is the mass of a helium atom).

At sufficiently low temperatures C_r must become smaller than C_{ph}. It must, however, be remarked that the calculated phonon heat capacity according to (3.1) is found to be at 1·2°K about 22 times less than its experimental value. In this way at temperatures even twice as low as the λ-point the roton gas still plays a predominant role in the heat capacity.

Up to the present time the heat capacity at lower temperatures has only been measured by B. Bleaney and F. Simon[6] (in a temperature interval from 0·25 to 0·8°K). Their results still give a value about 10 times larger than that given by the formula (3.1). On the other hand, Bleaney and Simon found it possible to describe the heat capacity in the temperature interval 0·25–0·8°K by a T^3-law. From this it would follow that the surplus part of the heat capacity over C_{ph}, i.e., the heat capacity C_r, also obeys the T^3-law, whereas according to the theory one would expect it to have an exponential temperature dependence. It must be mentioned, however, that Bleaney and Simon did not publish details of their measurements and so the possibility is not excluded that their results ought to be corrected. On the other hand, P. L. Kapitza's[9] measurements show that the value of the entropy at 1·33°K calculated from the formula (3.6) with μ and Δ from (3.8) agree well with experiment, whereas the entropy value calculated from the data of Bleaney and Simon does not agree with Kapitza's results.

4. SUPERFLUIDITY OF HELIUM II AT ABSOLUTE ZERO

It will now be shown that the superfluidity of helium II is a consequence of the properties of the energy spectrum described above. Consider first liquid helium at absolute zero. At this temperature the liquid is in its normal, unexcited state. Let us consider the helium when flowing through a capillary at a constant velocity V. The presence of viscosity would be manifested by the fact that owing to the friction between the liquid and the walls of the capillary and the friction inside the liquid itself the kinetic energy of the movement of the liquid would be dissipated and the flow gradually slowed down.

It is convenient to consider this flow in the coordinate system moving together with the helium. In this system the helium is at rest and the walls of the capillary move with a velocity $-V$. In the presence of viscosity the helium at rest should also begin to move. Physically it is obvious that the interaction of the liquid with the moving walls of the tube cannot cause a motion of the whole liquid at once. The motion must begin with the excitation, of the inner movements in the layers of liquid close to the walls of the tube, i.e., with the excitation of rotons and phonons in the liquid.

Let us suppose that a phonon can be excited in the liquid. Then the energy E_0 of the liquid (in the coordinate system in which it was initially at rest) is equal to the energy ε of the phonon (we will agree to take the energy of the normal state as zero) and the momentum $\boldsymbol{P}_0$ of the liquid is the momentum $\boldsymbol{p}$ of the phonon. As $\varepsilon = cp$ we have

$$E_0 = cp, \quad \boldsymbol{P}_0 = \boldsymbol{p}.$$

We will now return to the coordinate system in which the capillary is at rest. According to the known formulae for the transformation of energy and momentum in classical mechanics we have, for the energy E and the momentum $\boldsymbol{P}$ in this system

$$E = E_0 + (\boldsymbol{P}_0 \cdot \boldsymbol{V}) + \frac{MV^2}{2}, \quad \boldsymbol{P} = \boldsymbol{P}_0 + M\boldsymbol{V},$$

M being the mass of the liquid. Inserting E_0 and $\boldsymbol{P}_0$ we find for the energy

$$E = cp + (\boldsymbol{p} \cdot \boldsymbol{V}) + \frac{MV^2}{2}. \tag{4.1}$$

The term $MV^2/2$ is the initial kinetic energy of the flowing helium, the expression $cp + (\boldsymbol{p} \cdot \boldsymbol{V})$ being the change of energy owing to the excitation of the phonon. This change must be negative as the energy of the flowing liquid must decrease:

$$cp + (\boldsymbol{p} \cdot \boldsymbol{V}) < 0.$$

To fulfil this condition the absolute value of the velocity $\boldsymbol{V}$ must be, in any case, greater than the velocity of sound

$$V > c. \tag{4.2}$$

At smaller velocities the interaction with the walls of the tube cannot give rise to an excitation of phonons.

An analogous consideration can be applied to the excitation of a roton. In this case

$$E_0 = \varepsilon = \Delta + \frac{p^2}{2\mu},$$

and instead of (4.1) we get

$$E = \Delta + \frac{p^2}{2\mu} + (\boldsymbol{V} \cdot \boldsymbol{p}) + \frac{MV^2}{2}. \tag{4.3}$$

For $E < MV^2/2$ it is at any rate necessary that

$$\Delta + \frac{p^2}{2\mu} - Vp < 0,$$

$$V > \frac{p}{2\mu} + \frac{\Delta}{p}.$$

The expression on the right-hand side has a minimum at $p^2 = 2\mu\Delta$ so that for the excitation of rotons the velocity must be greater than

$$V > \sqrt{\frac{2\Delta}{\mu}}. \tag{4.4}$$

In this way we see that neither phonons nor rotons can be excited if the velocity of flow in helium II is not too large. This means that the flow of the liquid does not slow down, i.e., helium II discloses the phenomenon of superfluidity.*

It must be remarked that already the reasons given above are enough to make the superfluidity vanish at sufficiently large velocities. We leave aside the question as to whether superfluidity disappears at smaller velocities for some other reason [the velocity limit obtained from (4.2) is large—the velocity of sound in helium equals 250 m/sec; (4.4) gives a value only several times lower].

5. HELIUM II AT TEMPERATURES ABOVE ABSOLUTE ZERO

We now consider helium at non-zero temperatures. Helium is not then in its normal state and at sufficiently low temperatures its excitation can be considered as a gas consisting of phonons and rotons. The results of section 4 still remain valid as the circumstance that liquid helium is initially in the normal state was not used directly in our derivation. Therefore, new phonons and

* It must be mentioned that for an ideal gas $\Delta = 0$ and, therefore even at absolute zero it would not disclose the phenomenon of superfluidity at any velocities of the flow contrary to Tisza's suggestion.

rotons cannot be excited owing to the movement of the helium relative to the walls of the vessel also at temperatures greater than 0°K. It is necessary, however, to find out how the phonons and rotons that are already in the liquid will manifest themselves.

To investigate this problem we shall consider helium in an axial-symmetric vessel rotating round its axis at a constant angular velocity Ω. We go over to the coordinate system rotating together with the vessel. In this system the vessel is at rest, i.e., the external conditions in which the liquid is found are stationary. Therefore, the Gibbs distribution here holds, i.e., the probability of finding helium in an excited state is determined by the formula

$$\text{const } e^{-E'/kT},$$

E' being the energy of the excited state of the liquid in the rotating coordinate system. It is known that the energy E' of a body in the rotating coordinate system is connected with the energy in the stationary system by means of

$$E' = E - (\boldsymbol{M} \cdot \boldsymbol{\Omega}),$$

where $\boldsymbol{M}$ is the angular momentum of the body in a given excited state (in the normal state $\boldsymbol{M} = 0$). Thus, the Gibbs distribution in the stationary coordinate system is

$$\text{const } e^{-[E+(M\cdot\Omega)]/kT}. \tag{5.1}$$

For the sake of clarity let us suppose that the temperature is low enough so that one can speak of an ideal gas of phonons and rotons in helium. Then the energy E and angular momentum $\boldsymbol{M}$ of the excited state can be written as

$$E = \sum \varepsilon, \; \boldsymbol{M} = \sum \boldsymbol{m}, \tag{5.2}$$

ε and $\boldsymbol{m}$ being the energy and angular momentum respectively of single phonons and rotons.

As is well known, if we substitute $E = \sum \varepsilon$ in the Gibbs distribution $e^{-E/kT}$ we can pass over to the distribution for single "particles" of the gas, in the given case to the Bose distribution:

$$\frac{1}{e^{\varepsilon/kT}-1},$$

for phonons and probably for rotons (which particular distribution is not, however, essential to what will come).

Similarly, by inserting (5.2) into (5.1) we get in the same way the same distribution for phonons and rotons in the rotating vessel with the only difference that instead of ε enters $\varepsilon - (\boldsymbol{m} \cdot \boldsymbol{\Omega})$, so that the Bose distribution will take the form:

$$\frac{1}{e^{[\varepsilon-(\Omega \cdot m)]/kT}-1}.$$

But such a distribution is no other than the distribution for the system rotating as a whole (at an angular velocity $\boldsymbol{\Omega}$). And so we come to the result that during the rotation of the vessel containing helium II a statistical equilibrium is established which differs from the equilibrium in a vessel at rest only in that the photon and roton gas rotates together with the vessel as if it was carried along by the walls of the vessel.

If, from the above distribution one calculates the angular momentum of the helium in the rotating vessel at a given temperature, i.e., the quantity:

$$\boldsymbol{M} = \frac{1}{(2\pi\hbar)^3} \int\int \frac{\boldsymbol{m}\, d\tau_p\, dV}{e^{[\varepsilon-(\Omega \cdot m)]/kT}-1}, \tag{5.3}$$

then at absolute zero, i.e., when photons and rotons are completely absent we would obviously get zero. At higher temperatures the angular momentum will be non-zero but the momentum of inertia (i.e., the proportionality factor between $\boldsymbol{M}$ and $\boldsymbol{\Omega}$) will be at sufficiently low temperatures much lower than the usual one (which corresponds to the rotation of the total mass of the liquid with the vessel).*

* *(Note by Editor.)* Experiments did not confirm this picture (see, for example Osborne (1950) and Part 1 of the present volume).

This brings us to the fundamental result that when the walls of the vessel are in motion only a part of the liquid helium mass is carried along by them, the other part "remaining stationary". And so we might regard liquid helium as if it was a mixture of two liquids, one of which is superfluid having no viscosity and not carried along by the walls of the vessel and the other—a "normal" one which is carried along by the moving walls and behaves itself like a normal liquid. It is most essential that "there is no friction" between these two liquids moving through each other, that is, that there is no transfer of momentum from one to the other. Actually we get this relative motion when considering the statistical equilibrium in a uniformly rotating vessel. But if there can be some sort of relative motion in the state of statistical equilibrium it means that it cannot be accompanied by friction.

It must be stressed that when we look upon helium as a mixture of two liquids it is no more than a method of expression, convenient for describing phenomena which take place in helium II. Like every description of quantum phenomena in classical terms it is not quite adequate. Actually one must say that in a quantum liquid two movements can exist simultaneously, each of which is connected with its own "effective mass" (so that the sum of both these masses equals the total real mass of the liquid). One of these movements is normal, i.e., possesses the same properties as the movements of usual liquids; the other is superfluid. Both these motions take place without a transfer of the momentum from one to the other. We particularly emphasise that there is no division of the real particles of the liquid into "superfluid" and "normal" ones here. In a certain sense one can speak of "superfluid" and "normal" masses of liquid as of masses connected with the two simultaneously possible movements, but this by no means signifies the possibility of a real division of the liquid into two parts.

Keeping in mind these reservations as to the real nature of the phenomena taking place in helium II it is still convenient to use the terms "superfluid" and "normal" liquids as a short and convenient method of describing these phenomena and in the further discussion we shall do so.

During the rotation of the vessel containing helium the superfluid part remains stationary, as has already been pointed out. It can be said that the superfluid liquid is not capable of rotation. Mathematically this means that the curl of the velocity of the superfluid motion equals zero. Hence the motion of the superfluid liquid is always potential.* As to the normal part of the liquid it can accomplish a potential or a vortex motion.

This brings us in particular to the following interesting property of the motion of helium II. From hydrodynamics we know that during the potential flow of a liquid no pressure is exerted on a body immersed in the liquid (the so-called Euler paradox). Therefore, the superfluid part will not, during its motion, exert any pressure on a body immersed in helium II; the body will only be influenced by the normal part of the liquid.

At low temperatures, when the excitation is not too great the normal part of the liquid is an aggregate of phonons and rotons. Therefore, the viscosity of the normal liquid can be calculated in the same way as that of a mixture of phonon and roton gases. Even at not very low temperatures, i.e., at such temperatures when the number of rotons is large compared to the number of phonons, we cannot completely disregard the phonons and calculate the viscosity as the viscosity of the roton gas alone because the "mean free path" of phonons is, as can be shown, large as compared to that of the rotons.

The most important parameter which determines the properties of helium at every given temperature is the ratio of the masses of the superfluid and normal parts of the liquid. We introduce the density ϱ_n of the normal liquid and the density ϱ_s of the superfluid; the sum $\varrho = \varrho_n + \varrho_s$ is the total true density of the liquid.

* A more strict proof of this statement will be given in section 9, where the analogous questions in superconductors are being investigated.

At absolute zero the ratio ϱ_n/ϱ equals zero. As the temperature rises it increases until it becomes equal to unity, after which, of course, it will remain constant. The temperature at which ϱ_n/ϱ becomes unity is the point of transition from helium II into helium I. In this way the phase transition in helium is connected with the disappearance of the superfluid part of the liquid. This disappearance takes place continuously, i.e., ϱ_n/ϱ becomes unity continuously, without a jump. Therefore, the transition is a phase transition of the second order, i.e., a λ-point (it is not accompanied by the absorption or giving out of latent heat). The presence of a jump in the heat capacity is a direct thermodynamical consequence of a phase transition of the second order (cf., for example, ref. 7).

We will here point out the following possible method for a direct experimental determination of the curve of ϱ_n/ϱ as a function of the temperature. If the moment of inertia J of the vessel filled with helium II and rotating about its axis is measured the ratio of J to the moment of inertia J_0, calculated on the supposition that the total mass of helium rotates with the vessel, determines the ratio ϱ_n/ϱ at a given temperature.

At sufficiently low temperatures the ratio ϱ_n/ϱ can be calculated theoretically. At these temperatures ϱ is built up of two independent parts—the effective mass of an ideal gas of phonons and the mass of a roton gas. We will first calculate the phonon part of ϱ_n. To do this we shall imagine that the phonon gas moves as a whole with a constant velocity $\boldsymbol{V}$ relative to the liquid. As is known, the distribution function for a gas, moving as a whole, is obtained from the distribution function of the stationary gas simply by inserting the quantity $\varepsilon-(\boldsymbol{p}\cdot\boldsymbol{V})$ in place of the energy ε of the particle (where $\boldsymbol{p}$ is the momentum of the particle). Hence the distribution function for the phonons obeying Bose statistics is

$$\frac{1}{(2\pi\hbar)^3}\,\frac{1}{e^{[\varepsilon-(\boldsymbol{p}\cdot\boldsymbol{V})]/kT}-1}$$

We now calculate the total momentum $\boldsymbol{P}$ of the phonon gas (per unit volume). We have

$$\boldsymbol{P} = \frac{1}{(2\pi\hbar)^3} \int \frac{\boldsymbol{p}\, d\tau_p}{e^{[\varepsilon - (\boldsymbol{p}\cdot\boldsymbol{V})]/kT} - 1}.$$

Suppose that the velocity $\boldsymbol{V}$ is small and expand the integrated expression in powers of $(\boldsymbol{p}\cdot\boldsymbol{V})$. The zero order term vanishes after averaging over the directions of $\boldsymbol{p}$ and we obtain

$$\boldsymbol{P} = -\int \boldsymbol{p}(\boldsymbol{p}\cdot\boldsymbol{V}) \frac{\partial n_0}{\partial \varepsilon}\, d\tau_p,$$

$$n_0 = \frac{1}{(2\pi\hbar)^3} \frac{1}{e^{\varepsilon/kT} - 1}$$

As for the phonons we have $\varepsilon = cp$ and, therefore,

$$c\frac{\boldsymbol{p}}{p}\frac{\partial n_0}{\partial \varepsilon} = \frac{\partial n}{\partial \boldsymbol{p}},$$

so we can write

$$\boldsymbol{P} = -\frac{1}{c}\int \boldsymbol{p}(\boldsymbol{p}\cdot\boldsymbol{V})\frac{\partial n_0}{\partial \boldsymbol{p}}\, d\tau_p.$$

Integrating by parts and averaging over the directions of the momentum $\boldsymbol{p}$ we get

$$\boldsymbol{P} = \frac{4}{3c}\boldsymbol{V}\int p n_0\, d\tau_p = \frac{4}{3c^2}\boldsymbol{V}\int \varepsilon n_0\, d\tau_p.$$

But the integral entering this formula is the energy ϱE_{ph} per unit volume of the phonon gas, so that

$$\boldsymbol{P} = \frac{4}{3}\varrho \boldsymbol{V}\frac{E_{\text{ph}}}{c^2}.$$

The coefficient of $\boldsymbol{V}$ is the effective mass of a unit volume of the phonon gas, i.e., the phonon part $\varrho_n^{(\text{ph})}$ of the density ϱ_n:

$$\varrho_n^{(\text{ph})} = \frac{4}{3}\frac{E_{\text{ph}}}{c^2}\varrho. \tag{5.4}$$

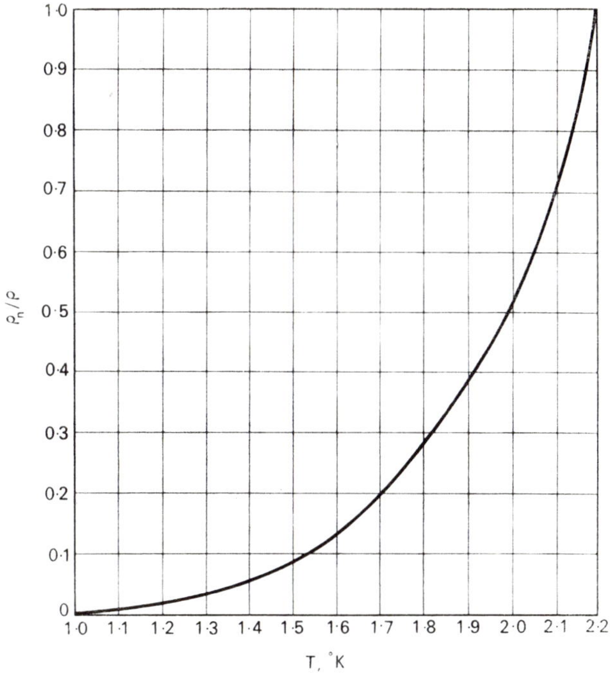

FIG. 1

The energy E_{ph} of the phonon gas, and hence $\varrho_n^{(ph)}$ is proportional to T^4.

A similar calculation gives the following expression for the roton part of the mass of the normal liquid at low temperatures (the Boltzmann distribution with $\varepsilon = \Delta + p^2/2\mu$ is used for the rotons):

$$\varrho_n^{(r)} = N_r \mu \varrho \tag{5.5}$$

[N_r is taken from (3.5).] The number N_r of rotons, and hence also $\varrho_n^{(r)}$ depend on the temperature exponentially ($\sim e^{-\Delta/kT}$, cf. section 3).

ϱ_n/ϱ is plotted against the temperature in Fig. 1. Even at 1°K the phonon part of ϱ_n is thirty times less than the roton part. The part of the curve in Fig. 1 which is near to the λ-point cannot, of course, be calculated theoretically and is obtained by means of an interpolation.

One may, however, expect that owing to a very rapid increase of ϱ_n according to formula (5.5), the value of the temperature of the λ-point itself can be approximately obtained from $\varrho_n/\varrho = 1$ using formula (5.5) for ϱ_n. Such a calculation gives 2·3°K for the temperature of the λ-point which is in a good agreement with the known value 2·19°K.

6. THE FLOW THROUGH CAPILLARIES AND THE HEAT CONDUCTIVITY OF HELIUM II

The ideas developed above allow one to give a satisfactory explanation of a number of results experimentally observed and also to predict a number of new properties which one might expect to find in helium II.

P. L. Kapitza[(1)] showed that helium II discloses no viscosity when flowing through a capillary or a narrow slit. From the point of view of the theory advanced here, this is explained by the fact that when helium II flows out of a vessel through a narrow slit the superfluid part of the liquid flows through the slit disclosing no friction; the normal part remains in the vessel flowing through the slit incomparably slower with a velocity corresponding to its viscosity and the thickness of the capillary.

It is known that when the viscosity is measured by means of observing the damping of the torsional vibrations of a disk immersed in helium II[(8)] one gets a non-zero value. This result also appears to be perfectly natural; actually, if the disk rotates in the liquid consisting of a superfluid and a normal part it will stop owing to the viscosity of the normal liquid. In this way, the experiment in which the liquid flows through a capillary discloses the presence of the superfluid part of the liquid and the experiment with the disk rotating in the helium II discloses the normal part.

The entropy of helium II is determined by the statistical distribution of the rotons and phonons. Hence for all motions of the liquid in which the phonon and roton gas remains stationary there is no macroscopical entropy transfer. This brings us to the most important result that during the flow of the superfluid liquid no entropy transfer takes place. In other words, the motion of the superfluid liquid does not carry heat. From this it follows in its turn that the motion of helium II in which only the superfluid part of the liquid takes part is thermodynamically reversible.

It was shown above that during the flow of helium II through a thin capillary or through a slit we have to deal with just such a flow of the superfluid liquid. Therefore, this flow must be reversible (to be more exact, the thinner the capillary and the less the normal liquid penetrates it the closer we are to a complete reversibility). This was disclosed during the recent ingenious experiments made by P. L. Kapitza[9].

As the helium flowing through a thin capillary does not carry heat, we may come to an important conclusion. Namely, liquid helium flowing out of a vessel through a thin capillary must be at a temperature of absolute zero (it would be more true to say at a temperature lower than the temperature of the helium in the vessel and equal to zero only in the ideal case of an infinitely thin capillary).

The known property of helium of forming a moving film on a hard surface is an effect akin to the properties of its flow through a thin capillary. The fact itself that a film is formed is not a peculiar property only to be found in helium II. Films are formed by all liquids wetting a given hard surface. But in ordinary liquids the formation of the film takes place extremely slowly owing to the presence of viscosity. The formation and movement of the film of helium II takes place quickly owing to the superfluidity.

The so-called thermomechanical effect in helium II is known to consist in the fact that when helium flows out of a vessel through a thin capillary a heating is observed in the vessel; on the other hand, at the place where the helium flows into a vessel from a capillary a cooling takes place. The presence of the thermomechanical effect is not in itself peculiar only to helium,* anomalous is only the large value of the effect. The thermomechanical effect in ordinary liquids is an irreversible phenomenon of the same type as the thermo-electric Peltier effect.

An effect of this type must also exist in helium II, however, in this case it is overlapped by a much stronger effect which is specific to helium II, having nothing in common with the irreversible phenomena of the type of the Peltier effect. Namely, the heating when the helium flows out through a capillary is simply due to the fact that this helium does not carry heat, therefore the heat in the vessel is distributed over a smaller amount of helium. When the helium flows in we find the reverse phenomenon.

It is easy to write what heat Q is absorbed when 1 g of helium flows out of the vessel through a capillary. As the liquid flowing out does not contain phonons and rotons its entropy equals zero. If the helium in the vessel is to remain at its initial temperature T the amount of heat TS (S—entropy of 1 g of helium at a temperature T) must be conveyed to it to compensate for the decrease which takes place per unit mass of the entropy owing to the introduction of 1 g of helium with zero entropy. This means that if 1 g of helium flows into a vessel containing helium at a temperature T, the amount of heat absorbed is

$$Q = TS. \tag{6.1}$$

In the opposite case when 1 g of helium flows out of the vessel containing helium at a temperature T this amount of heat is given off. The heat Q was recently measured by P. L. Kapitza[9]. The results obtained are found to be in excellent agreement with formula (6.1).

* The thermomechanical effect was observed recently in water by B. V. Derigauin.

We shall now consider two vessels containing helium II at temperatures T_1 and T_2 and connected to each other by a thin capillary. As the superfluid liquid is able to flow freely along the capillary a mechanical equilibrium of the helium in the two vessels is quickly established. But as, however, the superfluid liquid does not carry heat, the heat equilibrium at which the temperature of the helium in the two vessels equalises is only established much more slowly.

The conditions of the mechanical equilibrium of helium can be easily written down by using the fact that the establishment of this equilibrium takes place, according to what has gone before, at constant entropies S_1 and S_2 of helium in both vessels. If E_1 and E_2 are the internal energies per unit mass of helium at temperatures T_1 and T_2 the conditions of the mechanical equilibrium (i.e., the minimum energy) due to the transfer of the superfluid liquid is

$$\left(\frac{\partial E_1}{\partial N_1}\right)_{S_1} = \left(\frac{\partial E_2}{\partial N_2}\right)_{S_2},$$

N being the number of atoms in 1 g of helium. But the derivative $(\partial E/\partial N)_S$ is the chemical potential ζ. Therefore, the condition of equilibrium is obtained in the form

$$\zeta(p_1, T_1) = \zeta(p_2, T_2)$$

(p_1, T_1 and p_2, T_2 being the pressure and temperature in the first and second vessel) or

$$\Phi(p_1, T_1) = \Phi(p_2, T_2), \tag{6.2}$$

where Φ is the thermodynamic potential of 1 g of helium II.

If the pressures p_1 and p_2 are small, we can expand Φ in their powers and remembering that $\partial\Phi/\partial p$ is the specific volume V we get

$$V\Delta p = \Phi(T_1) - \Phi(T_2) = \int_{T_1}^{T_2} S\,dT \tag{6.3}$$

($\Delta p = p_2 - p_1$). If the temperature difference $\Delta T = T_2 - T_1$ is also small we can also expand in powers of ΔT and remarking that $\partial\Phi/\partial T = -S$ we get

$$\frac{\Delta p}{\Delta T} = \frac{S}{V} \tag{6.4}$$

As $S < 0$, and $V > 0$ also $dp/dT < 0$ in agreement with the experiment.

The formulae (6.1) and (6.4) were deduced already by H. London(10) starting from Tisza's ideas, the verbal formulation of which coincides at this point with the theory here advanced. These formulae are fully confirmed by P. L. Kapitza's(9) experiments.

We can note that the phenomena described can be regarded in the framework of the theory advanced here as the osmotic phenomena in the "solutions" of phonons and rotons in liquid helium, the narrow capillary or slit playing the role of a semipermeable membrane.

Finally, we come to the heat conductivity in helium II. The process of heat transfer in helium II must be represented in the following form. Consider the heat conductivity of helium II in a capillary the large value of which was discovered by W. Keesom and Miss Keesom(11) and studied in detail by P. L. Kapitza(12). If the helium is heated at one end of the capillary (or in a bulb soldered to it) two oppositely directed currents arise in helium. The normal liquid carrying the heat flows from the heated to the cold end; the heat which is transferred in this way is quite sufficient to explain the large value of the transfer experimentally observed. The current of the superfluid liquid flows in the opposite direction. The masses of liquid transferred by each current exactly compensate each other so that no real macroscopical flow in the helium takes place.

P. L. Kapitza(12) observed the deflection of a leaf hung in front of the open end of the capillary when the helium in the bulb at the other end of the capillary was heated. P. L. Kapitza suggested that the explanation of this phenomena might be that it was due to the axial flow of the helium from the heated to the cold end occupying practically the whole of the cross-section of the tube; the amount of helium in the tube was supposed to be sustained by means of the reverse flow of helium moving into the capillary along its surface.

According to the quantum representation developed above, this phenomenon has a different appearance. The current of superfluid liquid flowing into the capillary exerts no pressure on the leaf (owing to the potentiality of the movement, cf. section 5). On the other hand, the current of normal liquid exerts pressure on the leaf deflecting it in the direction observed. Both these currents go along the whole cross-section of the capillary and there is no more need to introduce the concept of a current of liquid flowing along the walls with anomalous properties.

It must again be emphasised that the concept of a superfluid and normal liquid is only a convenient way of describing the phenomena, but one must in reality speak of two movements which take place simultaneously in one and the same liquid; one of these movements carries heat and the other does not.

We thus come to a peculiar picture; on a body immersed in helium II a force can act in the absence of any movement of the liquid as a whole.

Consider also heat conductivity in large volumes of helium. If a body with a heated surface is immersed in helium II again two currents arise in the volume of liquid—the current of normal liquid flowing from the heat surface and an opposite current of the superfluid liquid. The transition of the currents one into the other, i.e., the transition of the superfluid liquid into the normal one takes place in the thin layer near the surface. The thickness of this layer must be of the order of the length of the free path of the phonons and rotons forming the normal part of the liquid. Almost the whole temperature fall will obviously take place in this layer and it follows that in the main volume of the liquid we can expect an almost total lack of temperature gradient. Such a distribution of temperature near the heated surface appears to be supported by P. L. Kapitza's[(12)] experiments.

7. EQUATIONS OF THE MACROSCOPIC HYDRODYNAMICS OF HELIUM II

Starting from the above considerations on the microscopic mechanism of the phenomenon of superfluidity a complete system of hydrodynamic equations can be built which would describe helium II in a macroscopic (phenomenological) way.

The starting point is the fundamental circumstance that the motion of helium II must be described not by one velocity as in ordinary hydrodynamics but by two. One of these is the "superfluid" velocity (denoted by $\boldsymbol{v}_s$) satisfying the condition

$$\text{curl}\, \boldsymbol{v}_s = 0. \tag{7.1}$$

On the boundary of a hard surface only the normal component of $\boldsymbol{v}_s$ becomes zero and not its tangential one corresponding to the fact that the superfluid liquid is not held back by friction against the walls of the vessel. For the "normal" velocity $\boldsymbol{v}_n$ of the liquid on the boundary with a hard surface the condition $\boldsymbol{v}_n = 0$ (as in ordinary viscous liquids) must be fulfilled which expresses the fact that the normal liquid is brought to a standstill owing to the friction against the walls.

It turns out that the hydrodynamic equations with the two velocities $\boldsymbol{v}_s$ and $\boldsymbol{v}_n$ can be obtained absolutely unambiguously starting from the one condition only that they should satisfy all the conservation laws. These equations for the general case of arbitrary velocities are somewhat complicated and we shall not give them here and confine ourselves to a simplified deduction of the equations applicable to the motion with not too large velocities $\boldsymbol{v}_s$ and $\boldsymbol{v}_n$.

Let $\boldsymbol{j}$ be the macroscopic current of the mass of liquid; it is a function of both the velocities $\boldsymbol{v}_s$ and $\boldsymbol{v}_n$. For small velocities $\boldsymbol{j}$ can be expanded in powers of $\boldsymbol{v}_s$ and $\boldsymbol{v}_n$. In the first approximation

$$\boldsymbol{j} = \varrho_s \boldsymbol{v}_s + \varrho_n \boldsymbol{v}_n. \tag{7.2}$$

The coefficients ϱ_s and ϱ_n are obviously those which we called the densities of the superfluid and normal "parts" of the liquid. Their sum equals the real density ϱ of helium II:

$$\varrho = \varrho_s + \varrho_n; \tag{7.3}$$

ϱ_s and ϱ_n are, of course, functions of temperature. Note that the current j (7.2) is at the same time the momentum density, i.e., the momentum of a unit volume of liquid. ϱ and $\boldsymbol{j}$ must satisfy the continuity equation:

$$\frac{\partial \varrho}{\partial t} + \operatorname{div} \boldsymbol{j} = 0. \tag{7.4}$$

We shall here write the equations applicable to a motion in which the viscosity of the "normal liquid" plays no part. Then the equation for the momentum conservation is written in the form

$$\frac{\partial j_i}{\partial t} + \frac{\partial \boldsymbol{\Pi}_{ik}}{\partial x_k} = 0 \tag{7.5}$$

(the summation is extended over the indexes which are repeated), where the tensor $\boldsymbol{\Pi}_{ik}$ of the momentum current equals

$$\boldsymbol{\Pi}_{ik} = p\delta_{ik} + \varrho_n v_i^{(n)} v_k^{(n)} + \varrho_s v_i^{(s)} v_k^{(s)}, \tag{7.6}$$

(p being the pressure). To take into account the viscosity of the normal liquid we must add to $\boldsymbol{\Pi}_{ik}$ the terms expressed in the ordinary way through the coefficients of viscosity and the derivatives of the velocity $\boldsymbol{v}_n$ with respect to the coordinates.

Further, the equation for the conservation of entropy takes the form

$$\frac{\partial S\varrho}{\partial t} + \operatorname{div}(\varrho S \boldsymbol{v}_n) = 0 \tag{7.7}$$

(S is the entropy per unit mass of helium II). The "entropy current" equals $\varrho S \boldsymbol{v}_n$, as the entropy is only transferred by the normal part of the liquid. If the viscosity of the normal part is taken into account, supplementary terms must be added to the right-hand side of (7.7) expressing the increase of the entropy owing to the irreversibility of the processes.

Finally, the last equation of the complete set of hydrodynamic equations we get by equalising the acceleration $d\boldsymbol{v}_s/dt$ to the force acting on a unit of the "superfluid" mass. To determine this force imagine that the unit mass of liquid is displaced from the point 1 to the point 2 in such a way that the distribution of phonons and rotons is not changed. In other words, one might say that during the transfer only the "superfluid liquid" is displaced and the distribution of the normal liquid remains unchanged. The energy E of the liquid changes during such a transfer by

$$\left(\frac{\partial E}{\partial M}\right)_1 - \left(\frac{\partial E}{\partial M}\right)_2$$

(M being the mass of the liquid). Derivatives must be taken here at constant entropy (because the entropy is connected only with the normal liquid) and at a constant momentum of the motion of the normal mass of the liquid relative to the superfluid;* besides this the volume of the liquid is considered as a constant.

From the expression obtained for the change of energy it is seen that the quantity $\partial E/\partial M$ can be regarded as a "potential energy" of the superfluid liquid, so that the force acting upon it is

$$-\operatorname{grad} \frac{\partial E}{\partial M}.$$

* The motion of the superfluid liquid may be considered as external conditions in which the phonons and rotons move. Therefore, the "Lagrange function" for the motion of the normal liquid does not simply depend on its velocity $\boldsymbol{v}_n$, but on the difference of the velocities $\boldsymbol{v}_n - \boldsymbol{v}_s$. The conserved momentum is, therefore, a derivative of the Lagrange function with respect to $\boldsymbol{v}_n - \boldsymbol{v}_s$, i.e., the momentum of the relative motion.

To calculate the derivative $\partial E/\partial M$ we notice that the derivative of the energy at constant entropy and volume is equal to the derivative of the thermodynamic potential at constant pressure and temperature. The thermodynamic potential $M\Phi$ of the liquid (Φ is the potential per unit mass) can be written in the form of the sum of the thermodynamic potential $M\Phi_0\,(p, T)$ of the stationary liquid and the kinetic energy $\boldsymbol{P}^2/2M_n$ of the relative motion of the superfluid and normal "parts":

$$M\Phi = M\Phi_0\,(p, T)+\frac{\boldsymbol{P}^2}{2M_n},$$

$\boldsymbol{P}$ is here the momentum of the motion of the normal mass M_n relative to the superfluid. By differentiating $M\Phi$ with respect to M at constant p, T and $\boldsymbol{P}$ and remembering that the normal mass M_n is proportional (at a given p and T) to the total mass M, we get

$$\Phi_0-\frac{\boldsymbol{P}^2}{2M_nM}.$$

If we insert $\boldsymbol{P} = M_n(\boldsymbol{v}_n-\boldsymbol{v}_s)$ and put the ratio of the densities in place of the ratio of the masses we finally find for the derivative $(\partial E/\partial M)_{S,V,P}$ the expression

$$\Phi_0-\frac{\varrho_n}{2(\varrho_n+\varrho_s)}(\boldsymbol{v}_n-\boldsymbol{v}_s)^2.$$

It follows that the hydrodynamic equation for which we were looking is of the form

$$\frac{d\boldsymbol{v}_s}{dt} = \frac{\partial \boldsymbol{v}_s}{\partial t}+(\boldsymbol{v}\cdot\nabla)\boldsymbol{v}_s = -\ \text{grad}\left\{\Phi-\frac{\varrho_n(\boldsymbol{v}_n-\boldsymbol{v}_s)^2}{2(\varrho_n+\varrho_s)^2}\right\}$$

(the index of Φ_0 is left out). It can be written differently if we use (7.1):

$$(\boldsymbol{v}_s\cdot\nabla)\boldsymbol{v}_s = \text{grad}\,\frac{\boldsymbol{v}_s^2}{2}.$$

In this way

$$\frac{\partial \boldsymbol{v}_s}{\partial t} = -\operatorname{grad}\left\{\Phi + \frac{v_s^2}{2} - \frac{\varrho_n(\boldsymbol{v}_n - \boldsymbol{v}_s)^2}{2(\varrho_n + \varrho_s)}\right\}. \tag{7.8}$$

The equations (7.1)–(7.8) are a complete set of hydrodynamic equations for helium II.

For a stationary flow the left-hand side of (7.8) is zero; hence:

$$\frac{v_s^2}{2} - \frac{\varrho_n(\boldsymbol{v}_n - \boldsymbol{v}_s)^2}{2(\varrho_n + \varrho_s)} + \Phi = \text{const.} \tag{7.9}$$

This equation together with the next one, (7.10), plays here the role of the Bernoulli equation.

Consider now the motions at which liquid may be considered incompressible. If we take the densities ϱ_n and ϱ_s and entropy S as constants we find from (7.4) and (7.7) that

$$\operatorname{div} \boldsymbol{v}_s = 0, \quad \operatorname{div} \boldsymbol{v}_n = 0.$$

Now, for a stationary motion we have in (7.5) $\partial \boldsymbol{j}/\partial t = 0$; by using

$$\frac{\partial v_k^{(s)}}{\partial x_k} \equiv \operatorname{div} \boldsymbol{v}_s = 0, \quad \frac{\partial v_k^{(n)}}{\partial x_k} = 0$$

we can rewrite (7.5) in the form

$$\nabla p + \varrho_n(\boldsymbol{v}_n \cdot \nabla)\boldsymbol{v}_n + \varrho_s(\boldsymbol{v}_s \cdot \nabla)\boldsymbol{v}_s = 0.$$

Remembering that curl $\boldsymbol{v}_s = 0$ we can write this equation as

$$\nabla\left(p + \varrho_n \frac{v_n^2}{2} + \varrho_s \frac{v_s^2}{2}\right) = \varrho_n[\boldsymbol{v}_n \wedge \operatorname{curl} \boldsymbol{v}_n].$$

We project this equation on the line of the current of the normal motion, i.e., on the direction of $\boldsymbol{v}_n$. Then on the right-hand side we get zero, so that

$$p + \varrho_n \frac{v_n^2}{2} + \varrho_s \frac{v_s^2}{2} = \text{const.} \tag{7.10}$$

It must be emphasised that the expression (7.10) is constant for a stationary flow only along each of the lines of current of the normal motion; and the expression (7.9) is constant over the whole volume of the liquid.

If the temperature and pressure change little over the volume of the liquid, Φ can be expanded in powers of $\Delta T = T - T_0$, $\Delta p = p - p_0$; T_0, p_0 being the temperature and pressure at a certain point in the liquid:

$$\Phi = \Phi(p_0, T_0) - S\,\Delta T + \frac{\Delta p}{\varrho}.$$

By inserting this into (7.9) we get

$$-\frac{v_s^2}{2} + \frac{\varrho_n(v_s - v_n)^2}{2\varrho} + S\,\Delta T - \frac{\Delta p}{\varrho} = \text{const.}$$

By combining this equation with equation (7.10)

$$\Delta p + \frac{\varrho_n v_n^2}{2} + \frac{\varrho_s v_s^2}{2} = \text{const.}$$

we get

$$\Delta T + \frac{\varrho_n}{\varrho S}(v_n \cdot v_n - v_s) = \text{const.} \tag{7.11}$$

This relation, like (7.10), is valid along the current lines of the normal motion.

8. PROPAGATION OF SOUND IN HELIUM II

The equations obtained can be applied to the propagation of sound in helium II. The velocity of the motion in sound waves is as usual supposed to be small and the density, pressure and entropy are almost equal to their constant equilibrium values. The terms in (7.6) and (7.8) which are quadratic with respect to the velocities can be neglected, and in (7.7) we can take the entropy ϱS in the term div $(v_n S\varrho)$ out of the sign of div as this term already contains the small quantity v_n. In this way the system of hydrodynamic equations for sound waves acquires the form

$$\frac{\partial \varrho}{\partial t}+\operatorname{div} \boldsymbol{j} = 0; \tag{8.1}$$

$$\frac{\partial \varrho S}{\partial t}+\varrho S \operatorname{div} \boldsymbol{v}_n = 0; \tag{8.2}$$

$$\frac{\partial \boldsymbol{j}}{\partial t}+\nabla p = 0; \tag{8.3}$$

$$\frac{\partial \boldsymbol{v}_s}{\partial t}+\nabla \Phi = 0. \tag{8.4}$$

By differentiating (8.1) with respect to time and inserting (8.3) we get

$$\frac{\partial^2 \varrho}{\partial t^2} = \nabla^2 p \tag{8.5}$$

Further, we have

$$\frac{\partial S}{\partial t} = \frac{1}{\varrho}\frac{\partial \varrho S}{\partial t}-\frac{S}{\varrho}\frac{\partial \varrho}{\partial t} = -S \operatorname{div} \boldsymbol{v}_n+\frac{S}{\varrho}\operatorname{div} \boldsymbol{j},$$

or

$$\frac{\partial S}{\partial t} = \frac{S\varrho_s}{\varrho}\operatorname{div}(\boldsymbol{v}_s-\boldsymbol{v}_n). \tag{8.6}$$

For the thermodynamic potential the relation

$$d\Phi = -SdT+Vdp = -SdT+\frac{1}{\varrho}dp$$

holds (V being the specific volume). Hence we have

$$\nabla p = S\varrho\nabla T+\varrho\nabla\Phi,$$

or by introducing ∇p from (8.3) and $\nabla\Phi$ from (8.4)

$$\varrho_n \frac{\partial}{\partial t}(\boldsymbol{v}_n-\boldsymbol{v}_s)+\varrho S\nabla T = 0. \tag{8.7}$$

Differentiating (8.6) with respect to time and introducing (8.7) we find

$$\frac{\partial^2 S}{\partial t^2} = \frac{S^2 \varrho_s}{\varrho_n} \nabla^2 T. \tag{8.8}$$

Two equations (8.5) and (8.8) determine the propagation of sound in helium II. It is already seen from the fact that there are two equations that there must be two velocities of sound in helium II.

Write S, ϱ, p, T in the form $S = S_0 + S'$, $\varrho = \varrho_0 + \varrho'$, etc. where the quantities with a dash represent the small changes of the corresponding quantities stipulated by the sound wave and the quantities with index zero are their constant equilibrium values. Then we can write

$$\varrho' = \frac{\partial \varrho}{\partial T} T' + \frac{\partial \varrho}{\partial p} p', \quad S' = \frac{\partial S}{\partial T} T' + \frac{\partial S}{\partial p} p',$$

and equations (8.5) and (8.8) take the form

$$\frac{\partial \varrho}{\partial p} \frac{\partial^2 p'}{\partial t^2} - \nabla^2 p' + \frac{\partial \varrho}{\partial T} \frac{\partial^2 T'}{\partial t^2} = 0,$$

$$\frac{\partial S}{\partial p} \frac{\partial^2 p'}{\partial t^2} + \frac{\partial S}{\partial T} \frac{\partial^2 T'}{\partial t^2} - \frac{S^2 \varrho_s}{\varrho_n} \nabla^2 T' = 0.$$

We look for a solution of these equations in the form of a plane wave in which p' and T' are proportional to a factor $e^{i\omega(t - x/u)}$ (u being the velocity of sound) and then for the conditions of solubility we get the equation

$$\begin{vmatrix} u^2 \dfrac{\partial p}{\partial \varrho} - 1 & u^2 \dfrac{\partial \varrho}{\partial T} \\ u^2 \dfrac{\partial S}{\partial p} & u^2 \dfrac{\partial S}{\partial T} - S^2 \dfrac{\varrho_s}{\varrho_n} \end{vmatrix} = 0$$

or

$$u^4 \frac{\partial(\varrho, S)}{\partial(p, T)} - u^2 \left(\frac{\partial S}{\partial T} + S^2 \frac{\varrho_s}{\varrho_n} \frac{\partial \varrho}{\partial \varrho} \right) + S^2 \frac{\varrho_s}{\varrho_n} = 0$$

[where $\partial(\varrho, S)/\partial(p, T)$ denotes the Jacobian of the transformation from ϱ, S to p, T]. By means of a simple transformation with the use of the thermodynamic relations this equation can be put in the form

$$u^4 - u^2\left[\left(\frac{\partial p}{\partial \varrho}\right)_S + \frac{TS^2}{C_v}\frac{\varrho_s}{\varrho_n}\right] + \frac{S^2\varrho_s T(\partial p/\partial \varrho)_T}{\varrho_s C_v} = 0 \qquad (8.9)$$

(C_v being the heat capacity of a unit mass of helium II). This quadratic equation determines two velocities of sound in helium II.

If $\varrho_s = 0$, i.e., at the λ-point, one of the roots of the equation (8.9) becomes zero and we get, as we ought, only one ordinary velocity of sound

$$u = \sqrt{\left(\frac{\partial p}{\partial \varrho}\right)_S}.$$

Practically for all temperatures the heat capacities C_p and C_v are close to each other. According to the known thermodynamic formula in these conditions the isothermic and adiabatic compressibilities are also close to each other, i.e.,

$$\left(\frac{\partial p}{\partial \varrho}\right)_S \approx \left(\frac{\partial p}{\partial \varrho}\right)_T.$$

If we denote the common value of $(\partial p/\partial \varrho)_T$ and $(\partial p/\partial \varrho)_S$ as $\partial p/\partial \varrho$ and the common value of C_p and C_v simply by C we get from the equation (8.9) two velocities of sound u_1 and u_2 in the form

$$u_1^2 = \frac{\partial p}{\partial \varrho}, \quad u_2^2 = \frac{TS^2}{C}\frac{\varrho_s}{\varrho_n}. \qquad (8.10)$$

In this way one of the velocities (u_1) is almost constant and the other (u_2) strongly depends on the temperature becoming zero at the λ-point. At a temperature 1·33°K we get a value of about 25 m/sec for u_2. At extremely low temperatures, $\varrho_n^{(ph)} \gg \varrho_n^{(r)}$ one gets

$$u_2 = \frac{c}{\sqrt{3}} \qquad (8.11)$$

In this way as the temperature tends to zero the velocity of sound tends to constant limits $u_1 = c$, $u_2 = c/\sqrt{3}$.

9. THE PROBLEM OF SUPERCONDUCTIVITY

The phenomenon of superconductivity is in many ways akin to the phenomenon of superfluidity. Superconductivity can also be explained by the supposition of the presence of an energy gap in the spectrum of the "electron liquid" in a metal. However, the character of this gap differs from that which is found in liquid helium. The raising of the question itself concerning the division of the motions of a liquid into potential and vortex is senseless when applied to the electron liquid in a metal.

The idea of a connection between superconductivity and the presence of a gap in the energy spectrum has been advanced several times [cf., for example, ref. 13]. However, there has been no clear proof given that the presence of this gap really leads to the phenomenon of superconductivity.

In the electron liquid in a metal one can distinguish between the motions of a pure inner character and the motions in which the liquid moves as a whole, in a macroscopic manner. The latter corresponds to the presence of a total current in the metal. The states with such a "macroscopic" motion begin directly from the normal unexcited state.* To explain superconductivity one must suppose that there is an energy interval between the normal state of the electron liquid and the beginning of the continuous spectrum of the states with an inner motion.

The value of this interval must be very small—in a temperature scale of the order of temperatures of the superconductive transition. What causes this gap is not, up to the present time, clear.

Let Ψ_0 be the wave function of the normal state of the electron liquid in which it is found at absolute zero. We shall consider a certain excited state which is close to the normal one. According to what was said this state is the state of the motion of the electron

* In this we have the essential distinction between superconductors and insulators in which there is also an energy gap, however, no states corresponding to a presence of a total current exist close to the normal state.

liquid as a whole. Corresponding to this the wave function Ψ of this state should have the form

$$\Psi = \Psi_0 \exp\left[\frac{i}{\hbar}(\boldsymbol{p}\cdot\sum_a \boldsymbol{r}_\alpha)\right]$$

where $\boldsymbol{p}$ is the momentum of the macroscopic motion per electron and the summation is extended over the radius-vectors of all the electrons.

Actually, however, one must keep in mind that a motion of the electron liquid as a whole must not necessarily have the same velocity in its whole volume. Therefore, in the wave function Ψ the exponential factor actually has the form exp $(i\sum\chi_\alpha/\hbar)$ and not exp $[i\boldsymbol{p}\cdot\sum\boldsymbol{r}_\alpha)/\hbar]$; χ is a certain function of the coordinates of one of the electrons about which one can only say that it is nearly linear; different terms in the sum $\sum\chi_\alpha$ differ in that χ is taken as a function of the coordinates $\boldsymbol{r}_\alpha$ of different electrons

$$\Psi = \Psi_0 \exp(i\sum\chi_\alpha/\hbar) \tag{9.1}$$

We determine the density $\boldsymbol{j}$ of the current in the electron liquid in the state Ψ. To do this we have to integrate the expression

$$\frac{\hbar e}{2im}(\Psi^*\nabla_\alpha\Psi - \Psi\nabla_\alpha\Psi^*) - \frac{e^2}{mc}A\Psi\Psi^*$$

(m being an "effective mass" of the electron, A—the vector potential of the magnetic field, e—the electron charge, c—the velocity of light) over the coordinates of all the electrons except the αth and then a summation must be made over all α (i.e., over all the electrons), taking the value of each term at the same point of the liquid. For $\Psi = \Psi_0$ and $A = 0$, i.e., for an electron liquid at rest, the current $\boldsymbol{j}$ obviously equals zero. The result is

$$\boldsymbol{j} = N\frac{e}{m}\,\text{grad}\,\chi\int|\Psi|^2\,d\tau' - N\frac{e}{mc}A\int|\Psi_0|^2\,d\tau'$$

($d\tau'$ is the product of the differentials of coordinates of all the electrons except of one of them; N is the total number of electrons).

But the integral

$$\int |\Psi|^2 \, d\tau' = \int |\Psi|^2 \, d\tau'$$

is the probability density of finding an electron at a given point of space. Therefore, the product $N \int |\Psi|^2 \, d\tau'$ is the number n of electrons in a unit volume of the electron liquid. In this way

$$\boldsymbol{j} = \frac{ne}{m} \operatorname{grad} \chi - \frac{e^2}{mc} n\boldsymbol{A}. \tag{9.2}$$

Eliminating χ we find

$$\operatorname{curl} \frac{mc\boldsymbol{j}}{e^2 n} = - \operatorname{curl} \boldsymbol{A} = -\boldsymbol{B} \tag{9.3}$$

($\boldsymbol{B} = \operatorname{curl} \boldsymbol{A}$ is the magnetic induction). This equation may be regarded as a supplementary condition which must be satisfied in the possible states of the macroscopic motion of the electron liquid. It is to be noted that this relation is a generalisation of the equation $\operatorname{curl} \boldsymbol{v}_s = 0$ which is found for superfluidity and goes over into it if the magnetic field can be neglected.

If the metal is homogeneous the electron density can be considered constant and the equation (9.3) can also be written as

$$\operatorname{curl} \boldsymbol{j} = -\frac{ne^2}{mc} \boldsymbol{B}. \tag{9.4}$$

From this equation [first put forward by R. Becker, G. Heller, and F. Sauter(14) and first written in the form (9.4) by London(15)] the magnetic (Meissner effect) and the electric (lack of resistance) properties of superconductors follow. It is known also that (9.4) is also supported quantitatively by the experiments on the superconductivity of helium films(16).

Similarly to the conclusions in section 5 one can come to the conclusion that at non-zero temperatures the equation (9.4) still holds with the only difference that en now equals not the total charge density of the electron liquid but only a certain portion of it. In this way one can in a certain sense speak of the division of the whole charge density into "superconducting" and "normal" parts, but not forgetting all the reservations mentioned for the similar division of the density of the mass of liquid helium. In particular, this does not mean that there are "superconducting" and "normal" electrons in metals; simply the motion of the electron liquid in the metal can be described as a combination of two simultaneous motions each of which is connected with a certain effective charge. The superconductivity vanishes at that temperature at which the "superconducting" charge becomes zero.

It must be noted, however, that in the equation (9.4) the charge en enters in a combination ne/m with the mass m. Therefore, only that ratio has a physical sense and not each one of the values of en and m separately.

As in helium II we come to the conclusion that the superconducting current must not transfer heat. This is supported by the well-known fact that the thermoelectric phenomena are absent in superconductors[17].

In conclusion, I wish to express my thanks to P. L. Kapitza for kindly informing me of his experimental results before publication.

REFERENCES

1. P. L. Kapitza, *Nature*, **141**, 74 (1938); *Collected Papers*, Pergamon, 1965, Vol. II, ch. 34.
2. L. Tisza, *Nature*, **141**, 913 (1938).
3. A. Bijl, *Physica*, **7**, 869 (1940).
4. W. H. Keesom, *Leiden Comm., Supp.*, 80b (1936).
5. W. H. Keesom and A. P. Keesom, *Physica*, **2**, 557 (1935).
6. B. Bleaney and F. Simon, *Trans. Faraday Soc.*, **35**, 1205 (1939).
7. L. Landau and E. Lifshitz, *Statistical Physics* (Oxford, 1938).
8. W. Keesom and G. McWood, *Physica*, **5**, 257 (1938).
9. P. L. Kapitza, *J. Phys.*, **5**, 59 (1941); *Collected Papers*, Pergamon, 1965, Vol. II, ch. 39.
10. H. London, *Proc. Roy. Soc.*, **A 171**, 484 (1939).
11. W. H. Keesom and A. P. Keesom, *Physica*, **3**, 359 (1936).
12. P. L. Kapitza, *J. Phys.*, **4**, 181 (1941); *Collected Papers*, Pergamon, 1965, Vol. II, ch. 38.
13. H. Welker, *Z. Phys.*, **114**, 535 (1939).

14. R. Becker, G. Heller and F. Sauter, *Z. Phys.*, **85**, 772 (1933).
15. F. and H. London, *Proc. Roy. Soc.*, **A 149**, 71 (1935).
16. E. T. S. Appleyard and J. T. Bristow, *Proc. Roy. Soc.*, **A 172**, 530 (1939); A. Shalnikov, *Nature*, **142**, 74 (1938).
17. W. H. Keesom and G. J. Matthijs, *Physica*, **5**, 437 (1938).

T2. RADIATION OF SOUND IN HELIUM II*

E. LIFSHITZ

THE microscopic theory of the phenomenon of superfluidity of helium II was developed by L. Landau[(1)]. The same author, on the basis of these ideas, developed the hydrodynamics of helium II describing its macroscopic motion. In this hydrodynamics the motion of the liquid is described simultaneously by two velocities: a velocity of "superfluid motion" v_s and a velocity of "normal motion" v_n.

Upon application of the system of hydrodynamical equations to the propagation of sound in helium II [ref. 1, § 8], it appears that the propagation of sound waves in this medium can take place with two different velocities, which in the first approximation are equal to

$$u_1 = \sqrt{\frac{\partial p}{\partial \varrho}}, \quad u_2 = \sqrt{\frac{TS^2}{C}\frac{\varrho_s}{\varrho_n}} \tag{1}$$

(S, C are the entropy and heat capacity per unit mass of liquid, ϱ_s and ϱ_n—the densities of the "superfluid" and "normal" parts of the liquid). In this approximation terms have been neglected containing a small quantity—the coefficient of thermal expansion; in connection with this no distinction is made between $(\partial p/\partial \varrho)_T$ and $(\partial p/\partial \varrho)_S$ and between the heat capacities C_p and C_v. The velocity u_1 corresponds to the ordinary velocity of sound, it depends slightly on the temperature and is almost constant over

* *J. Phys. (U.S.S.R.)* **8**, 110 (1944).

the whole range from 0°K to the λ-point; at 1·5–2°K $u_1 = 2{\cdot}4 \cdot 10^4$ cm/sec. On the other hand, the velocity u_2 changes rapidly with the temperature; it becomes zero together with ϱ_s at the λ-point and increases monotonously to the value $u_1/\sqrt{3}$ at 0°K. Fig. 1 gives the graph of the function $u_2(T)$ in the interval from 0·8°K to the λ-point, calculated from the formulae for $C, S, \varrho_s/\varrho_n$ obtained by L. Landau.

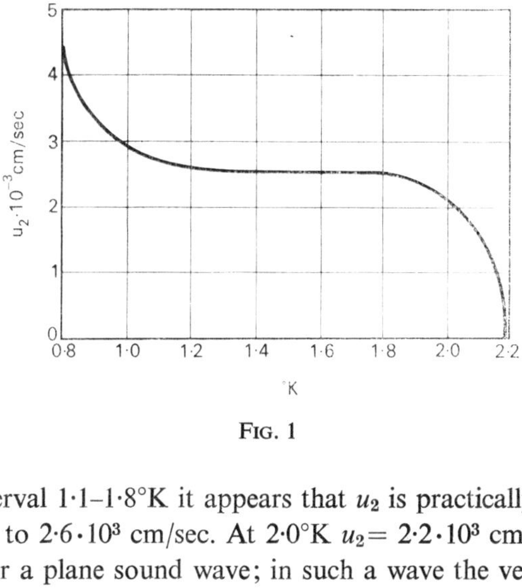

FIG. 1

In the interval 1·1–1·8°K it appears that u_2 is practically constant and equal to $2{\cdot}6 \cdot 10^3$ cm/sec. At 2·0°K $u_2 = 2{\cdot}2 \cdot 10^3$ cm/sec.

Consider a plane sound wave; in such a wave the velocities v_s, v_n and the variable parts T', p' of the temperature and pressure are proportional to each other. Upon introducing proportionality factors a, b, c according to

$$v_n = av_s, \quad p' = bv_s, \quad T' = cv_s, \tag{2}$$

and making simple calculations with the aid of equations (8.1–8) in ref. 1 to a sufficient degree of approximation we obtain for the "first sound":

$$a_1 = 1 + \frac{\alpha\varrho}{S\varrho_s} \frac{u_1^2 u_2^2}{(u_1^2 - u_2^2)}, \quad b_1 = \varrho u_1, \quad c_1 = \frac{\alpha T}{C} \frac{u_1^3}{(u_1^2 - u_2^2)}, \tag{3}$$

and for the "second sound":

$$a_2 = -\frac{\varrho_s}{\varrho_n} + \frac{\alpha\varrho}{S\varrho_n}\frac{u_1^2 u_2^2}{(u_1^2-u_2^2)}, \quad b_2 = \frac{\alpha\varrho u_1^2 u_2^3}{S(u_1^2-u_2^2)}, \quad c_2 = -\frac{u_2}{S} \tag{4}$$

($\alpha = -\frac{1}{\varrho}\frac{\partial\varrho}{\partial T}$ is the coefficient of heat expansion). It should be observed that in sound waves of the first type $v_s \cong v_n$, i.e. in such a wave the liquid vibrates as a whole in each volume element; these waves correspond to ordinary sound waves in ordinary liquids. In a wave of the second type $v_n \cong (-\varrho_s/\varrho_n)v_s$, i.e. the total flow of the mass of liquid, $\boldsymbol{j} = \varrho_s \boldsymbol{v}_s + \varrho_n \boldsymbol{v}_n \sim 0$; thus, here the superfluid and normal components of the liquid vibrate with respect to each other in such a manner that in the first approximation their centre of "inertia" in each volume element remains stationary and there is no total flow of matter. We also observe that in waves of this type the amplitude of vibration of the pressure is small (p' is proportional to α) and that of the temperature—large, whereas in the "first sound" the opposite is true.

The character of the sound radiation depends essentially on the ratio of the dimensions of the radiating body to the wave length. Consider first the case when the wave length (of both sounds) is small compared with the dimensions of the vibrating body. It can be assumed then that each part of the body surface radiates a plane wave in a direction normal to itself. It will therefore be sufficient to consider the radiation of sound by a plane surface vibrating as a whole in a direction perpendicular to itself (which we shall choose for the direction of the x axis). We look for the velocity v_s (along the x axis) in the "first" and "second" sound waves in the form

$$v_s^{(1)} = A_1 e^{-i\omega\left(t-\frac{x}{u_1}\right)}, \quad v_s^{(2)} = A_2 e^{-i\omega\left(t-\frac{x}{u_2}\right)}, \tag{5}$$

respectively. The boundary conditions at the surface of a solid body at rest require, in general, that the tangential component of velocity $\boldsymbol{v}_n$ and the normal component of the total flow of matter

$\boldsymbol{j} = \varrho_s \boldsymbol{v}_s + \varrho_n \boldsymbol{v}_n$ disappear; the normal component of heat flow in helium II $\varrho ST \boldsymbol{v}_n$ must be equal to the heat flow in a solid body due to thermal conduction(2). However, due to the high rate of heat transmission in helium II compared with that in a solid, we can neglect the latter. The last two boundary conditions then reduce to the disappearance of the normal components of each of the velocities $\boldsymbol{v}_s$ and $\boldsymbol{v}_n$, or—on a moving surface—to their equality to the corresponding component of velocity of the body.

In the present case the boundary conditions give

$$A_1 + A_2 = u_0, \quad A_1 a_1 + A_2 a_2 = u_0$$

(where $u_0 e^{-i\omega t}$ is the velocity of vibration of the solid surface) whence

$$A_1 = u_0 \frac{1-a_2}{a_1-a_2}, \quad A_2 = -u_0 \frac{1-a_1}{a_1-a_2}. \tag{6}$$

As in all small vibrations the average (over the time) potential energy equals the average kinetic energy. Hence the total energy density of a sound wave in helium II equals $\varrho_s v_s^2 + \varrho_n v_n^2 = \frac{|A|^2}{2}(\varrho_s + \varrho_n a^2)$; the flow of energy is obtained by multiplying by the corresponding velocity of sound u. From the expressions for A_1, A_2 and the values of a_1, a_2 we thus obtain for the ratio of the intensities I_1, I_2 of the radiated waves of the "first" and "second" sounds:

$$\frac{I_2}{I_1} = \frac{\alpha^2 T}{C} \frac{u_2^3 u_1^3}{(u_1^2 - u_2^2)^2} \cong \frac{a^2 T}{C} \frac{u_2^3}{u_1} \tag{7}$$

(down to the very lowest temperatures $u_2 \ll u_1$, which is assumed, for brevity, in all that follows). For the ratios of the amplitudes of pressure and temperature in both waves we obtain

$$\frac{p_2'}{p_1'} = \frac{\alpha^2 T}{C} \frac{u_2^3}{u_1}, \quad \frac{T_2'}{T_1'} = \frac{u_2}{u_1}. \tag{8}$$

Thus, at $T = 2{\cdot}0°K$ the ratio I_2/I_1 is equal to $2 \cdot 10^{-6}$; hence under these conditions (radiation of sound of small wave length) the "second sound" is practically not radiated at all. The ratio of the pressure amplitudes is just as small, while that of the temperature amplitudes is more favourable to the second sound, being equal at the same temperature to 0·1.

Consider now the sound radiation of a sphere performing translatory vibrations, the radius R of the sphere being small in comparison with the radiated wave lengths (since $u_2 < u_1$ this reduces to the condition $R \ll u_2/\omega$). At distances from the sphere r large in comparison with R, we look for the sound velocities $\boldsymbol{v}_s$ in both waves in the form:

$$\boldsymbol{v}_s^{(1)} = A_1(\boldsymbol{u}\nabla)\nabla\frac{e^{i\frac{\omega}{u_1}r}}{r}, \quad \boldsymbol{v}_s^{(2)} = A_2(\boldsymbol{u}\nabla)\nabla\frac{e^{i\frac{\omega}{u_2}r}}{r} \tag{9}$$

(where $\boldsymbol{u} = \boldsymbol{u}_0 e^{-i\omega t}$ is the velocity of the sphere). At distances large also in comparison with the wave lengths λ_1, λ_2 the factor $1/r$ must be considered as a constant when differentiating. In this region (the wave zone) the wave may be regarded as plane in every small region; correspondingly the velocities $\boldsymbol{v}_n^{(1)}$, $\boldsymbol{v}_n^{(2)}$ here must be equal to $\boldsymbol{v}_n^{(1)} = a_1\boldsymbol{v}_s^{(1)}$, $\boldsymbol{v}_n^{(2)} = a_2\boldsymbol{v}_s^{(2)}$. On going over to distances large in comparison to R, but small in comparison to the wave lengths, expressions (9) become

$$\boldsymbol{v}_s^{(1)} = A_1(\boldsymbol{u}\nabla)\nabla\frac{1}{r}, \quad \boldsymbol{v}_s^{(2)} = A_2(\boldsymbol{u}\nabla)\nabla\frac{1}{r}.$$

On the other hand, at distances $r \ll \lambda$ the liquid may be regarded as incompressible and the distribution of the velocity $\boldsymbol{v}_s = \boldsymbol{v}_s^{(1)} + \boldsymbol{v}_s^{(2)}$ of the superfluid motion must be determined by the usual formula for the potential flow of an ideal incompressible liquid past a sphere, i.e. it should be $\boldsymbol{v}_s = \frac{R^3}{2}(\boldsymbol{u}\nabla)\nabla\frac{1}{r}$. From this follows the condition

$$A_1 + A_2 = \frac{R^3}{2}.$$

As for the distribution of the velocity $\boldsymbol{v}_n = \boldsymbol{v}_n^{(1)} + \boldsymbol{v}_n^{(2)}$ of normal flow, at distances $r \ll \lambda$, it should be determined by the solution of the Navier–Stokes equation (Stokes flow about a vibrating sphere). The rôle of the kinematic viscosity ν is played here by the ratio η/ϱ_n (η being the viscosity of helium II)(2). At distances $R \ll r \ll \lambda$ the distribution has the form [see, for instance(3), p. 643]

$$v_n = \frac{R^3}{2}\left(1+\frac{3(1+i)}{2\varkappa}+\frac{3i}{2\varkappa^2}\right)(\boldsymbol{u}\nabla)\nabla\frac{1}{r} \quad \left(\varkappa = R\sqrt{\frac{\omega}{2\nu}}\right)$$

(here terms have been omitted which decrease exponentially with r). From this we obtain the second condition in the form:

$$A_1a_1 + A_2a_2 = \frac{R^3}{2}\left(1+\frac{3(1+i)}{2\varkappa}+\frac{3i}{2\varkappa^3}\right).$$

These two equalities yield

$$A_1 = \frac{R^3}{2}\left\{1+\frac{3(1+i)}{2\varkappa}\frac{\varrho_n}{\varrho}+\frac{3i}{2\varkappa^2}\frac{\varrho_n}{\varrho}\right\},$$

$$A_2 = \frac{R^3}{2}\left\{\frac{\alpha\varrho_n}{S\varrho_s}u_2^2-\frac{3(1+i)}{2\varkappa}\frac{\varrho_n}{\varrho}-\frac{3i}{2\varkappa^2}\frac{\varrho_n}{\varrho}\right\}. \qquad (10)$$

In the "wave zone" we have for each of the velocities $\boldsymbol{v}_s^{(1)}$, $\boldsymbol{v}_s^{(2)}$, $\boldsymbol{v}_n^{(1)}$, $\boldsymbol{v}_n^{(2)}$ an expression of the form

$$v = A(\boldsymbol{u}\nabla)\nabla\frac{e^{i(\omega/u)r}}{r} = -A\frac{\omega^2}{u^2}\frac{\boldsymbol{n}(\boldsymbol{n}\cdot\boldsymbol{u})}{r}e^{i(\omega/u)r}$$

($\boldsymbol{u}$ is a unit vector in the direction of $\boldsymbol{r}$). Averaging of v^2 over the time gives

$$|A|^2\frac{\omega^4}{6u^4}\frac{|\boldsymbol{u}_0|^2}{r^2}.$$

Hence, the total intensity of the "first sound" is

$$I_1 = 4\pi r^2 u_1(\varrho_s\overline{v_s^{(1)2}}+\varrho_n\overline{v_n^{(1)2}}) = \frac{2\pi}{3}\frac{\omega^4}{u_1^3}|u_0|^2|A_1|^2(\varrho_s+\varrho_n a_1^2),$$

and a similar expression for I_2. For the ratio of the two intensities we get the following expression:

$$\frac{I_2}{I_1} = \frac{u_1^3}{u_2^3}\frac{\varrho_s}{\varrho_n}\frac{\left(\frac{\alpha\varrho}{S\varrho_s}u_2^2-\frac{3}{2\varkappa}\right)^2+\frac{9}{4\varkappa^2}\left(1+\frac{1}{\varkappa}\right)^3}{\left(\frac{\varrho}{\varrho_n}+\frac{3}{2\varkappa}\right)^2+\frac{9}{4\varkappa^2}\left(1+\frac{1}{\varkappa}\right)^2}. \tag{11}$$

In the limiting case of high frequencies this yields

$$\frac{I_2}{I_1} = \frac{\alpha^2 T}{C}\frac{u_1^3}{u_2}, \tag{12}$$

while in the opposite case of low frequencies

$$\frac{I_2}{I_1} = \frac{u_1^3}{u_2^3}\frac{\varrho_s}{\varrho_n}. \tag{13}$$

Thus, at $T = 2°K$, $R = 0{\cdot}1$ cm, $\omega/2\pi = 50\ \text{sec}^{-1}$, we obtain $I_2/I_1 = 0{\cdot}2$ [we have taken for η the value $2\cdot10^{-5}$ poise, cf.(2).] The limiting values of (12) and (13) at the same temperature are equal respectively to $2{\cdot}5\cdot10^{-3}$ and $1{\cdot}4\cdot10^{3}$. Thus, with regard to total intensity, the radiation of sound by a sphere vibrating at low frequencies may be favourable for the "second sound".

It is difficult, however, to observe the intensity of sound directly. It might be possible, for example, to observe a wave by the vibrations of another sphere or a plate which it sets into motion. Calculation shows, however, that the absolute values of the amplitudes of vibration which might be expected here are so insignificant as to be practically unobservable.

There are still other means of exciting sound waves in helium II which are specific for that medium, but impossible in ordinary liquids. Consider the radiation of sound by a plate vibrating in its own plane*; let the dimensions of the plate be small as compared

* This method of sound radiation was suggested by P. L. Kapitza.

with the wave length. At large distances we may once more look for the velocities in the form (9). At distances where the liquid can be regarded as incompressible, only its "normal" part is drawn along by the plate, so that the velocity of the superfluid motion must be equal to zero. This gives us one of the boundary conditions in the form: $A_1+A_2=0$, so that $|A_2/A_1|=1$. Corresponding to this, we obtain for the ratio of the intensities expression (13). The ratio of the amplitudes of the temperature vibrations equals

$$\frac{T_3}{T_2}=\frac{C}{\alpha TS}\,\frac{u_2}{u_1} \tag{14}$$

(at 2°K this ratio is equal to 20). Radiation from a surface whose temperature fluctuates periodically in time is even more favourable for the "second sound". Consider a plane solid surface whose temperature varies according to the law

$$T'=T_0'e^{-i\omega t}.$$

The conditions to be required at this surface, in accordance with the exact boundary conditions [see[2]], are the disappearance of the normal component of the total flow of matter and equality of the temperatures of liquid and wall. As for the flow of heat from the wall to the liquid, it is determined from the derived solution as the product $\varrho ST v_n$. We look for the solution in the form (5) and, bearing in mind that a change of temperature in a sound wave is connected with v_s by means of the relation $T'=cv_s$ [where c is defined by formulae (3), (4)], write the boundary conditions in the form

$$\varrho_s(A_1+A_2)+\varrho_n(a_1A_1+a_2A_2)=0;$$
$$c_1A_1+c_2A_2=T_0'.$$

For the ratio A_2/A_1, which interests us, we have

$$\left|\frac{A_2}{A_1}\right|=\frac{\varrho_s+a_1\varrho_n}{\varrho_s+a_2\varrho_n}\cong\frac{S}{\alpha u_2^2}.$$

Hence we find the ratio of intensities in the form

$$\frac{I_2}{I_1} = \frac{C}{T\alpha^2 u_1 u_2}. \tag{15}$$

At $T = 2°K$ this quantity is equal to $5 \cdot 10^3$ while at lower temperatures it is even greater. Thus, as regards intensity, practically only the "second sound" is radiated in this case. For the ratios of the amplitudes of vibration of pressure and temperature we obtain

$$\frac{p_2'}{p_1'} = \frac{u_2}{u_1}, \quad \frac{T_2'}{T_1'} = \frac{C}{\alpha^2 T u_1 u_2}. \tag{16}$$

At $T = 2°K$, $p_2'/p_1' = 0{\cdot}1$, so that the conditions for observing the "second sound" by the variation in pressure are still unfavourable. On the contrary, the conditions for observing the "second sound" by the variation in temperature are extremely favourable.

The following general note should be added. As is well known, the property of superfluidity disappears at velocities exceeding a certain critical value, which depends on the temperature, as well as on the characteristic dimensions of the problem. For large dimensions of the vessel containing the helium II the critical velocity is rather small, probably of the order of several centimetres per second. For the unhampered observation of the second sound it is necessary that the velocity of motion of liquid in the sound wave would not exceed the critical values (at 2°K $T' = 2 \cdot 10^{-4}$ corresponds to a velocity $v = 1$ cm/sec.).

In conclusion I wish to express my gratitude to Prof. L. Landau for his interest in the work and for valuable discussions.

REFERENCES

1. L. Landau, *Journ. of Phys.*, **5**, 71 (1941) See p. 191 of the present book.
2. L. Landau, *Journ. of Phys.*, **8**, 1 (1944).
3. H. Lamb, *Hydrodynamics*, 6th ed., Cambridge, 1931.

T3. ON THE THEORY OF SUPERFLUIDITY OF HELIUM II*

L. D. LANDAU

THE velocity of the "second sound" in helium II has been measured by V. Peshkov[1] with a great precision. His results give an opportunity to perform a quantitative comparison of the theory developed by the author[2] with experiment. Such a comparison gives full support to the general picture given by the theory, but at the same time it reveals a noticeable discrepancy between the calculated and observed values of the velocity (e.g., 25 m/sec calculated and 19 m/sec observed at a temperature of 1·6°K). Although this discrepancy is not very large, it is too large to be attributed to the inaccuracy of the experimental data on the thermodynamic quantities of helium II.

For calculating the velocity of the second sound the formulae were used for the thermodynamic quantities (entropy, specific heat), derived in ref. 2 under the assumption of the energy spectrum of the liquid to consist of two branches—the phonon and roton ones. The direction of the observed discrepancy indicates in what way these assumptions must be altered. Using the experimental data, one can formally compute the roton mass μ according to formulae

$$\varrho_n = N\mu, \ F_r = -NkT. \tag{1}$$

Here N is the number of rotons per unit volume, F_r—the "roton part" of the free energy (i.e., the free energy without the vibrational

* *J. Phys. (U.S.S.R.)* **11**, 91, 1947.

part), ϱ_n—the density of the "normal part" of the liquid (the phonon part in ϱ_n is negligible as compared with the roton part). The mass μ calculated in this way appears to be approximately inversely proportional to the temperature (in the temperature interval 1·3–1·7°K), instead of being constant. It is, however, to be noted, that although the very fact of the variation of μ is apparent, the quantitative law of its variation can be established only in a very

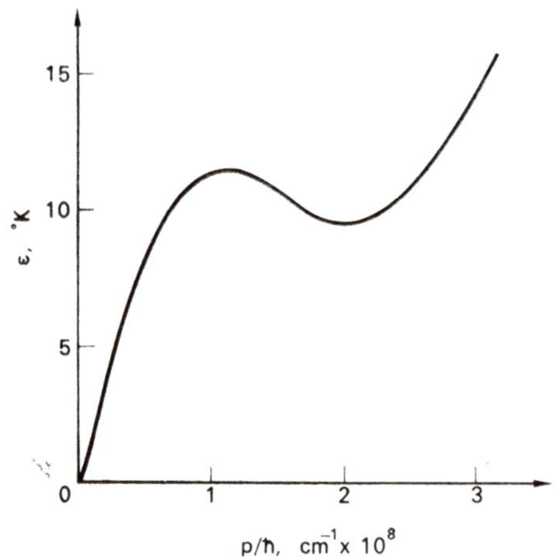

FIG. 1

approximate way (owing mainly to the scarcity of experimental data on the specific heat of helium II).

If one does not make the assumption $\varepsilon = \Delta + p^2/2\mu$ for the dependence of the energy ε of a roton on its momentum p, but considers the general dependence $\varepsilon(p)$, then the calculation according to the general formulae derived in ref. 2 shows, that in the formula $\varrho_n = N\mu\varrho$ the quantity $\overline{p^2}/3kT$ enters instead of μ ($\overline{p^2}$ is the mean square of the momentum). If this quantity is inversely proportional to the temperature, then $\overline{p^2}$ = const, i.e., the values of the roton momenta lie mainly in the neighbourhood of a certain p_0.

At the first glance this fact appears to be very strange, but it can be explained in a natural manner by assuming that the energy spectrum of helium II is of the type shown in Fig. 1. For small momenta p of an elementary excitation its energy ε increases linearly (phonons), then reaches a maximum, begins to decrease and at a certain value $p = p_0$ the function $\varepsilon(p)$ has a minimum. In the neighbourhood of the latter we can write

$$\varepsilon = \Delta + \frac{(p-p_0)^2}{2\mu}. \tag{2}$$

μ being a constant. With such a spectrum it is of course impossible to speak strictly of rotons and phonons as of qualitatively different types of elementary excitations. It would be more correct to speak simply of the long wave (small p) and short wave (p in the neighbourhood of p_0) excitations.* It is to be stressed, that all the conclusions concerning the superfluidity and the entire macroscopical hydrodynamics of helium II, developed in ref. 2, maintain their validity also with the spectrum proposed here.

Only the formulae for the thermodynamic quantities must be changed. Instead of formulae (3, 4–7) in ref. 2 we have for the "roton" parts of the free energy, entropy, specific heat (per unit mass) and the density of the "normal liquid":

$$F_r = -\frac{2\mu^{1/2}(kT)^{3/2}p_0^2}{(2\pi)^{3/2}\varrho\hbar^3}e^{-\Delta k/T}; \tag{3}$$

$$S_r = \frac{2(k\mu)^{1/2}p_0^2\Delta}{(2\pi)^{3/2}\varrho T^{1/2}\hbar^3}\left(1+\frac{3kT}{2\Delta}\right)e^{-\Delta/kT}; \tag{4}$$

$$C_r = \frac{2\mu^{1/2}p_0^2\Delta^2}{(2\pi)^{3/2}\varrho k^{1/2}T^{3/2}\hbar^3}\left[1+\frac{kT}{\Delta}+\frac{3}{4}\left(\frac{kT}{\Delta}\right)^2\right]e^{-\Delta/kT}; \tag{5}$$

$$\frac{(\varrho_n)_r}{\varrho} = \frac{2\mu^{1/2}p_0^4}{3(2\pi)^{3/2}\varrho(kT)^{1/2}\hbar^3}e^{-\Delta/kT}. \tag{6}$$

* *(Note by Editor.)* See Bogolubov's paper reprinted as paper T4 in this volume.

In such a form the theory contains three constants: Δ, p_0 and μ. It is, therefore, difficult to check it on the basis of the experimental data which are now available. For the values of Δ, p_0 and μ one gets*

$$\frac{\Delta}{k} = 9\ 6^{\circ}, \quad \frac{p_0}{\hbar} = 1{\cdot}95\times 10^{8}\,\mathrm{cm}^{-1}, \quad \mu = 0{\cdot}77\, m_{\mathrm{He}}. \tag{7}$$

Note that μ is of the order of the mass m_{He} of the helium atom and $\hbar/p_0$ is even less than the atomic dimensions. The values (7) have been used in drawing the curve in Fig. 1.

REFERENCES

1. V. PESHKOV, *J. Phys.*, **10,** 389 (1946).
2. L. LANDAU, *J. Phys.*, **5,** 71 (1941); See p. 191 of the present book.

* *(Note by Editor.)* The values of the constants can be also found from neutron-scattering experiments (Yarnell, Arnold, Bendt and Kerr (1959), Henshaw and Woods (1961)).

T4. ON THE THEORY OF SUPERFLUIDITY*

N. Bogolubov

Summary. This paper presents an attempt of explaining the phenomenon of superfluidity on the basis of the theory of degeneracy of a nonperfect Bose–Einstein gas.

By using the method of the second quantization together with an approximation procedure we show that in the case of the small interaction between molecules the low excited states of the gas can be described as a perfect Bose–Einstein gas of certain "quasi-particles" representing the elementary excitations, which cannot be identified with the individual molecules.

The special form of the energy of a quasi-particle as a function of its momentum is shown to be connected with the superfluidity.

The object of this paper is an attempt to construct a consistent molecular theory explaining the phenomenon of superfluidity without assumptions concerning the structure of the energy spectrum.

The most natural starting point for such a theory seems to be the scheme of a nonperfect Bose–Einstein gas with a weak interaction between its particles.

It should be noted that similar attempts were done some time ago by Tisza and London to explain the phenomenon of superfluidity on the basis of the degeneracy of a perfect Bose–Einstein gas, but these attempts raised a counterblast of objections.

It has been pointed out, for example, that helium II has nothing to do with a perfect gas, because of the strong interaction between

* *J. Phys. (USSR)* **11**, 23 (1947).

its molecules. However, this objection cannot be regarded as an essential one. Indeed, it is clear that a rigorous theoretical computation of the properties of a real liquid is hopelessly beyond the reach of a pure molecular theory based on usual "microscopic" equations of quantum mechanics. All we can require from a molecular theory of superfluidity, at least at the first stage of investigation, is to be able to account for the qualitative picture of this phenomenon being based on a certain simplified scheme.

A really essential objection one can make against this idea is the following one. The particles of a degenerate perfect Bose–Einstein gas in the ground state cannot possess the property of superfluidity, since nothing prevents them from exchanging their momenta with excited particles colliding with them, and, therefore, from friction in their movement through the liquid.

In the present paper we try to overcome this difficulty and to show that under certain conditions the "degenerate condensate" of a "nearly perfect" Bose–Einstein gas can move without any friction with respect to the elementary excitations, with an arbitrary, sufficiently small velocity. It is to be pointed out that the necessity of considering the collective elementary excitations rather than individual molecules was suggested by L. Landau in his well-known paper "Theory of Superfluidity of Helium II" where he, by postulating their existence in form of phonons and rotons, was enabled to explain the property of the superfluidity.

In our theory the existence and the properties of the elementary excitations follow directly from the basic equations describing the Bose–Einstein condensation of non perfect gases.

1. Let us consider a system of N identical monoatomic molecules enclosed in a certain macroscopic volume V and subjected to Bose statistics.

Suppose the Hamiltonian of our system to be, as usually, of the form

$$H = \sum_{(i \leq i \leq N)} T(p_i) + \sum_{(i \leq i < j \leq N)} \Phi(|q_i - q_j|),$$

where

$$T(p_i) = \frac{|p_i|^2}{2m} = \sum_{(i\leq\alpha\leq 3)} \frac{(p_i^\alpha)^2}{2m}$$

represents the kinetic energy of the ith molecule and $\Phi(|q_i - q_j|)$—the mutual potential energy of the pair (i, j).

Applying then the method of second quantization, let us present the basic equation in the form

$$i\hbar\frac{\partial\Psi}{\partial t} = -\frac{\hbar^2}{2m}\nabla^2\Psi + \int \Phi(|q-q'|)\,\Psi^+(q')\,\Psi(q')\,dq'\cdot\Psi, \quad (1)$$

whereby

$$\Psi = \sum_f \alpha_f\varphi_f(q); \quad \Psi^+ = \sum_f \alpha_f^+\varphi_f^+(q).$$

Here a_f, a_f^+ are conjugated operators with commutation rules of the well-known type

$$a_f a_{f'} - a_{f'} a_f = 0;$$

$$a_f a_{f'}^+ - a_{f'}^+ a_f = \Delta_{f,f'} = \begin{cases} 1, & f = f' \\ 0, & f \neq f' \end{cases}$$

and $\{\varphi_f(q)\}$ is a complete orthonormal set of functions

$$\int \varphi_f^+(q)\varphi_{f'}(q)dq = \Delta_{f,f'}.$$

For the sake of simplicity, we shall further employ the system of eigenfunctions of the momentum operator of a single particle

$$\varphi_f(q) = \frac{1}{V^{1/2}} e^{i\frac{(f\cdot q)}{\hbar}}; \; (f\cdot q) = \sum_{(i\leq a\leq 3)} f^\alpha q^\alpha,$$

the operator $N_f = a_f^+ a_f$ then representing the number of molecules with momentum f. For finite V vector f is obviously quantized. For example, under usual boundary conditions of the periodicity type,

$$f^\alpha = \frac{2\pi n^\alpha \hbar}{l},$$

where (n^1, n^2, n^3) are integers, and l denotes the side of a cube of the volume V.

However, since we are going to deal with thermodynamic volume properties of the system, we have to bear in mind a limiting process, such that when $N \to \infty$, the boundary of the container tends to infinity, $V \to \infty$, but the volume per molecule $v = V/N$ remain constant. Therefore, we shall pass over in the final results to a continuous spectrum, replacing the sums $\sum_f F(f)$ by integrals

$$\frac{V}{(2\pi\hbar)^3} \int F(f)\, df.$$

Equations (1) are the exact equations of the problem of N bodies. It becomes, therefore, a necessity in order to push forward the investigation of the motion of the considered system, to apply some approximation method based on a supposition that the energy of interaction is sufficiently small. According to this supposition, the potential function $\Phi(r)$ is assumed to be proportional to a certain small parameter ε.

Later on we shall see what dimensionless quantity can be chosen for ε. For the present it suffices to notice that the above supposition, strictly speaking, corresponds to a neglection of the finiteness of molecular radius, since we do not take into account the intensive increase of $\Phi(r)$ for small r, which causes the impenetrability of molecules. We shall see, however, that the results, which will be obtained, can be generalized so as to include the case of a finite molecular radius.

Now let us turn to the formulation of an approximate method. If there be no interaction at all, i.e. if ε be exactly zero, we could put at zero temperature: $N_0 = N$, $N_f = 0\ (f \neq 0)$.

But in the case under consideration, when ε is small and the gas is in a weakly excited state, these relations are valid approximately, which means that the momenta of the overwhelming majority of molecules approach zero. Of course, the fact that zero momentum is the limiting one for particles in the ground state, is due to a specific choice of the coordinate system: namely, we choose the system with respect to which our "condensate" is at rest.

Our approximation method, based on the above considerations, runs as follows.

(1) Since $N_0 = a_0^+ a_0$ is very large as compared with unity, the expression

$$a_0 a_0^+ - a_0^+ a_0 = 1,$$

is small as compared with a_0^+, a_0 themselves, which enables us to treat them as ordinary numbers* neglecting their non-commutability

(2) Let us put

$$\Psi = \frac{a_0}{\sqrt{V}} + \vartheta; \quad \vartheta = \frac{1}{\sqrt{V}} \sum_{(f \neq 0)} a_f e^{i \frac{(f \cdot q)}{\hbar}}$$

and consider ϑ as a "correction term of the first order". Neglecting all the terms in equation (1), involving the second and higher powers of ϑ, this being permissible since the excitations are supposed to be weak, we obtain the following basic approximate equations:

$$\begin{aligned} i\hbar \frac{\partial \vartheta}{\partial t} = & -\frac{\hbar^2}{2m} \nabla^2 \vartheta + \frac{N_0}{V} \Phi_0 \vartheta + \\ & + \frac{N_0}{V} \int \Phi(|q-q'|) \vartheta(q') \, dq' + \\ & + \frac{a_0^2}{V} \int \Phi(|q-q'|) \vartheta^+(q') \, dq', \end{aligned} \qquad (2)$$

$$i\hbar \frac{\partial a_0}{\partial t} = \frac{N_0}{V} \Phi_0 a_0,$$

where

$$\Phi_0 = \int \Phi(|q|) \, dq.$$

* A similar procedure was used by Dirac in his book *The Principles of Quantum Mechanics* (second edition), cf. the end of § 63: Waves and Bose–Einstein particles.

In order to pass from the operator wave function ϑ to operator amplitudes let us apply Fourier's expansion

$$\Phi(|q-q'|) = \sum_f \frac{1}{V} e^{i\frac{(f\cdot q-q')}{\hbar}} \nu(f). \tag{3}$$

From the radial symmetry of the potential function it follows that the amplitudes of this expansion

$$\nu(f) = \int \Phi(|q|) e^{-i\frac{(f\cdot q)}{\hbar}} dq,$$

depend upon the length $|f|$ of vector f only.

The substitution of (3) into equation (2) gives

$$i\hbar \frac{\partial a_f}{\partial t} = \left\{T(f)+E_0+\frac{N_0}{V}\nu(f)\right\} a_f + \\ +\frac{a_0^2}{V}\nu(f)a_{-f}^{+};$$

$$E_0 = \frac{N_0}{V}\Phi,$$

whence on setting

$$a_f = e^{\frac{E_0}{i\hbar}t} b_f; \quad a_0 = e^{\frac{E_0}{i\hbar}t} b \tag{4}$$

we get

$$i\hbar \frac{\partial b_f}{\partial t} = \left\{T(f)+\frac{N_0}{V}\nu(f)\right\} b_f + \frac{b^2}{V}\nu(f)b_{-f}^{+},$$

$$-i\hbar \frac{\partial b_{-f}^{+}}{\partial t} = \frac{(b^{+})^2}{V}\nu(f)b_f + \left\{T(f)+\frac{N_0}{V}\nu(f)\right\} b_{-f}^{+}.$$

Solving this system of two differential equations with constant coefficients, we find that the operators b_f, b_f^{+} depend upon the time by means of a linear combination of exponentials possessing the form

$$e^{\pm i\frac{E(f)}{\hbar}t},$$

where*

$$E(f) = \sqrt{2T(f)\frac{N_0}{V}\nu(f)+T^2(f)}. \tag{6}$$

Now let us observe that the inequality

$$\nu(0) = \int \Phi(|q|)\,dq > 0, \tag{7}$$

implies the positiveness of the expression under the sign of the radical (6) since ε is considered to be sufficiently small; thus b_f, b_f^+ prove to be periodical functions of time. On the contrary, if $\nu(0) < 0$, this expression is negative for small momenta and, therefore, $E(f)$ receives complex values. As a consequence, b_f, b_f^+ will involve a real exponential increasing with time, whence it follows that the states with small $N_f = b_f^+ b_f$ are unstable.

In order to be sure in the stability of the excited states, let us restrict the class of possible types of interaction forces, supposing inequality (7) to be satisfied for all types we shall consider. It is interesting to note that the inequality (7) just represents the condition of thermodynamic stability of a gas at absolute zero.

Indeed, at absolute zero the free energy coincides with the mean energy. The main term of the latter has the form

$$E = \frac{N^2}{2V}\int \Phi(|q|)\,dq,$$

since the correction terms (for instance, the mean kinetic energy) are proportional to higher powers of ε.

Therefore, the pressure P is expressed by the following formula:

$$P = -\frac{\partial E}{\partial V} = \frac{N^2}{2V^2}\int \Phi(|q|)\,dq = \frac{\varrho^2}{2m^2}\,\nu(0),$$

where $\varrho = Nm/V$ represents the density of the gas.

* We may remark that if we form the corresponding frequency $\omega = \hbar E(f)$ and put: $\hbar \to 0, f/h = \hbar =$ const we obtain the classical formula for the dispersion of the frequency deduced by A. Vlasov, cf. *Journal of Physics*, **9**, 25 (1945).

This proves that the inequality (7) is equivalent to the condition of thermodynamic stability

$$\frac{\partial P}{\partial \varrho} > 0.$$

Finally, let us note that we can write instead of (6), with the same degree of accuracy

$$E(f) = \sqrt{2T(f)\frac{N}{V}\nu(f)+T^2(f)} = \sqrt{\frac{|f|^2\nu(f)}{mv}+\frac{|f|^4}{4m^2}}, \quad (6')$$

since we take into account the main terms only; it follows from (6′) for small momenta

$$E(f) = \sqrt{\frac{\nu(0)}{mv}}\,|f|\,(1+\ldots) = \sqrt{\frac{\partial P}{\partial \varrho}}\,|f|\,(1+\ldots),$$

where by dots are denoted the terms vanishing together with f.

In what follows we shall attribute the positive sign to each square root we shall deal with. We thus have for small momenta

$$E(f) = c\,|f|\,(1+\ldots), \quad (8)$$

where c denotes the velocity of sound at absolute zero.

On the contrary, for sufficiently large momenta $E(f)$ can be expanded in powers of ε

$$E(f) = \frac{|f|^2}{2m}+\frac{\nu(f)}{v}+\ldots \quad (9)$$

Since $\nu(f)$ tends to zero with increasing $|f|$,* $E(f)$ is seen to approach the kinetic energy $T(f)$ of a single molecule for sufficiently large momenta.

Returning now to equations (5), let us introduce new mutually conjugated operators ξ_f, ξ_f^+ instead of b_f, b_f^+ by means of the

* *(Note by Editor)*. Thus, $d\nu(f)/df < 0$ which allows us to have $dE(f)/df = 0$ and therefore for the intermediate momenta the simplest form of the energy-momentum curve is that assumed by Landau.

following relations:

$$\xi_f = \frac{b_f - L_f b^+_{-f}}{\sqrt{1-|L_f|^2}}; \quad \xi_f^+ = \frac{b_f^+ - L_f^+ b_{-f}}{\sqrt{1-|L_f|^2}}, \tag{10}$$

where L_f are numbers determined by the equalities

$$L_f = \frac{Vb^2}{N_0^2 \nu(f)} \left\{ E(f) - T(f) - \frac{N_0}{V} \nu(f) \right\},$$

so that

$$|L|^2 = \left(\frac{(V_0/V)\, \nu(f)}{E(f) + T(f) + (N_0/V)\nu(f)} \right)^3;$$

$$1 - |L_f|^2 = \frac{2E(f)}{E(f)+T(f)+(N_0/V)\nu(f)}. \tag{11}$$

Reversing (10) we obtain:*

$$b_f = \frac{\xi_f + L_f \xi^+_{-f}}{\sqrt{1-|L|^2}}; \quad b_f^+ = \frac{\xi_f^+ + L_f^+ \xi - f}{\sqrt{1-|L_f|^2}} \tag{12}$$

The substitution of these expressions into equations (5) gives

$$i\hbar \frac{\partial \xi_f}{\partial t} = E(f)\xi_f; \quad -i\hbar \frac{\partial \xi_f^+}{\partial t} = E(f)\, \xi_f^+. \tag{13}$$

It can be immediately verified that the new operators satisfy the same commutation relations as the operators a_f, a_f^+:

$$\xi_f \xi_{f'} - \xi_{f'} \xi_f = 0; \quad \xi_f \xi_{f'}^+ - \xi_{f'}^+ \xi_f = \Delta_{f,f'}. \tag{14}$$

This alone suffices to conclude that the excited states of the given assemblage of molecules can be treated as a perfect gas composed of "elementary excitations"—"quasi-particles" with energy depending on the momentum by means of the relation: $E = E(f)$. These quasi-particles are described by the operators ξ_f, ξ_f^+ in the same way as molecules were described by the operators a_f, a_f^+, and,

* *(Note by Editor)*. This transformation was, with great success, applied by N. Bogolubov to the theory of superconductivity.

therefore, they are also subjected to Bose statistics. The operator

$$n_f = \xi_f^+ \xi_f$$

represents the number of quasi-particles with the momentum f.

The above conclusion becomes quite clear if we consider the total energy

$$H = H_{\text{kin}} + H_{\text{pot}}$$

where

$$H_{\text{kin}} = -\frac{\hbar^2}{2m} \int \Psi^+(q) \nabla^2 \Psi(q)\, dq,$$

$$H_{\text{pot}} = \frac{1}{2} \int \Phi(|q-q'|) \Psi^+(q) \Psi^+(q') \Psi(q) \Psi(q')\, dq\, dq' =$$

$$= \frac{1}{2V} \sum_f \nu(f) \int e^{\frac{(f \cdot q - q')}{\hbar}} \Psi^+(q) \Psi^+(q') \Psi(q) \Psi(q')\, dq\, dq'.$$

For the kinetic energy we obtain

$$H_{\text{kin}} = \sum_f T(f) a_f^+ a_f = \sum_f T(f) b_f^+ b_f.$$

In order to compute the potential energy with the assumed degree of accuracy, we shall disregard in the expression

$$\Psi^+(q) \Psi^+(q') \Psi(q) \Psi(q') =$$

$$= \left(\frac{a_0^+}{\sqrt{V}} + \vartheta^+(q')\right) \left(\frac{a_0^+}{\sqrt{V}} + \vartheta^+(q')\right) \left(\frac{a_0}{\sqrt{V}} + \vartheta(q)\right) \left(\frac{a_0}{\sqrt{V}} + \vartheta(q')\right)$$

all terms of the third and higher order with respect to ϑ, ϑ^+. This gives

$$H_{\text{pot}} = \Phi_0 \left\{ \frac{1}{2} \frac{N_0^2}{V} + \frac{N_0}{V} \sum_{f \neq 0} b_f^+ b_f \right\} + \frac{b^2}{2V} \sum_{f \neq 0} \nu(f) b_f^+ b_{-f}^+ +$$

$$+ \frac{(b^+)^2}{2V} \sum_{\neq 0} \nu(f) b_f b_{-f} + \frac{N_0}{V} \sum_{f \neq 0} \nu(f) b_f^+ b_f.$$

Noticing here that

$$\sum_{f \neq 0} b_f^+ b_f = \sum_{f \neq 0} N_f = N - N_0,$$

we can write with the same degree of accuracy

$$\frac{1}{2}\,\frac{N_0^2}{V}+\frac{N_0}{V}\sum_{f\neq 0} b_f^+ b_f=\frac{1}{2}\,\frac{N^2}{V}\,.$$

Hence

$$H=\frac{N^2}{2V}\,\Phi_0+\frac{b^2}{2V}\sum_{f\neq 0}\nu(f)b_f^+b_{-f}^{+}+\frac{(b^+)^2}{2V}\sum_{f\neq 0}\nu(f)b_f b_{-f}+$$

$$+\frac{N_0}{V}\sum_{f\neq 0}\nu(f)b_f^+b_f+\sum_f T(f)b_f^+b_f\,.$$

Replacing here the operators b_f, b_f^+ by the operators ξ_f, ξ_f^+ with the aid of relations (12), we obtain finally

$$H=H_0+\sum_{f\neq 0}E(f)\,n_f;\quad n_f=\xi_f^+\xi_f, \tag{15}$$

where

$$H_0=\frac{1}{2}\,\frac{N^2}{V}\,\Phi_0+\sum_{f\neq 0}\frac{E(f)-T(f)-(N_0/V)\,\nu(f)}{2}=$$

$$=\frac{1}{2}\,\frac{N^2}{V}\,\Phi_0+\frac{V}{2(2\pi\hbar)^3}\int\left\{E(f)-T(f)-\frac{N_0}{V}\,\nu(f)\right\}df. \tag{16}$$

Thus we see that the total energy of the considered non-ideal gas consists of the energy of the ground state and the individual energies of each of the quasi-particles. It means that the quasi-particles do not interact with each other and thus form a perfect Bose–Einstein gas. The absence of interaction between the quasi-particles is evidently caused by the admitted approximation, namely by neglecting the terms of the third and higher order with respect to ξ_f, ξ_f^+ involved in the expression for energy. Therefore, the above results are valid for weakly excited states only.

On having taken into account the disregarded third-order terms in the expression for energy or the second-order terms in equations (13) which are to be considered as a small perturbation, we could discover a weak interaction between the quasi-particles. This interaction enables the assemblage of quasi-particles to attain the state of statistical equilibrium. Proceeding to the study of this state, let us first observe that the total momentum of quasi-particles: $\sum_f f n_f$, is conserved. To prove this consider the components of the total momentum. From

$$\sum_{(i\leq i\leq N)} p_i^\alpha = \int \Psi^+(q)\left\{-i\hbar\frac{\partial\Psi(q)}{\partial q^\alpha}\right\}dq = \sum_f f^\alpha a_f^+ a_f = \sum_f f^\alpha b_f^+ b_f$$

it follows, according to the transformation relations (12), that

$$\sum_{(1\leq i\leq N)} p_i^\alpha = \sum_f f^\alpha \frac{(\xi_f^+ + L_f^+\xi_{-f})(\xi_f + L_f\xi_{-f}^+)}{1-|L_f|^2}.$$

The invariance of L_f, L_f^+ with respect to the replacement of f by $-f$ implies

$$\sum_f f^\alpha \frac{L_f^+\xi_{-f}\xi_f}{1-|L_f|^2} = \sum_f f^\alpha \frac{L_f^+\xi_f^+\xi_{-f}^+}{1-|L_f|^2} = 0,$$

$$\sum_f f^\alpha \frac{|L_f|^2\xi_{-f}\xi_{-f}^+}{1-|L_f|^2} = \sum_f f^\alpha \frac{|L_f|^2(\xi_{-f}\xi_{-f}^+ - 1)}{1-|L_f|^2} = -\sum_f f^\alpha \frac{|L_f|^2\xi_f^+\xi_f}{1-|L_f|^2}.$$

Hence we conclude that

$$\sum_{(i\leq i\leq N)} p_i^\alpha = \sum_f f^\alpha n_f,$$

i.e. that the total momentum of the assemblage of molecules is equal to that of the assemblage of quasi-particles. Since the former is conserved, the sum $\sum_f f n_f$ proves to be conserved too.

It is also easy to see that the total number of quasi-particles $\sum_f n_f$ is not invariant; quasi-particles can appear and disappear.

For this reason we obtain in the usual way the following formula for average occupation numbers $\bar{n}_f (f\neq 0)$ in the state of statistical equilibrium

$$\bar{n}_f = \left\{A\exp\frac{E(f)-(f\cdot u)}{\theta} - 1\right\}^{-1}; \quad A = 1, \tag{17}$$

where θ is the temperature modulus, while u denotes an arbitrary vector. However, the length of this vector must have an upper limit. In fact, since all average occupation numbers have to be positive, the inequality

$$E(f) > (f \cdot u),$$

and, therefore, the inequality

$$E(f) > |f| \cdot |u|$$

is satisfied for all $f \neq 0$.

But from the above properties of $E(f)$ it follows that the ratio

$$\frac{E(f)}{|f|}$$

is a continuous positive function of $|f|$. This function is equal to $c > 0$ at $|f| = 0$ and is increasing as $|f|^2/2m$, for $|f| \to \infty$; therefore, the minimum value of the considered ratio is essentially positive. Hence the condition for the positiveness of $\bar{n}_f$ is equivalent to inequality

$$|u| \leqslant \min \frac{E(f)}{|f|}. \tag{18}$$

If the decrease of $E(f)$ for small momenta be proportional not to the momentum itself, but to its square (as it is the case for the kinetic energy of a molecule), the right-hand side of the obtained inequality would be equal to zero, this value being the only possible one for u. But in the case under consideration the vector u can be chosen arbitrarily, provided that its length is sufficiently small.

Now let us note that from the momentum distribution over the gas, composed of quasi-particles, given by formula (17), it follows that this gas is moving as a whole with the velocity u. At first, we have chosen such a coordinate system with respect to which the condensate (i.e. the assemblage of molecules in the ground state) is at rest. Inversely, by transition to a coordinate system with respect to which the quasi-particle gas, as a whole, is at rest, we can discover the motion of the condensate with the velocity u.

This relative motion goes on stationarily in the state of statistical equilibrium without any external forces. Hence we see that it is not accompanied by friction and thus represents the property of superfluidity*.

As we have seen, the energy of a quasi-particle is asymptotically equal to $c|f|$ for small momenta, c being the velocity of sound. Therefore, a quasi-particle for small momenta is just a phonon. When the momentum is increasing, the kinetic energy becomes large as compared with the binding energy of a molecule, and the energy of a quasi-particle tends continuously toward the individual energy of a molecule $T(f)$.

Thus we see that no division of quasi-particles into two different types, phonons and rotons, can even be spoken of.

2. Now let us consider the distribution of momenta over an assemblage of molecules, in the state of statistical equilibrium. We introduce a function $W(f)$ defined in such a way that $NW(f)\,df$ represents the average number of molecules whose momenta belong to an elementary volume df of the momentum space. This function is seen to be normalized by means of

$$\int W(f)\,df = 1. \tag{19}$$

Suppose, further, $F(f)$ to be an arbitrary continuous function of the momentum. For the mean value of a dynamical variable

$$\sum_{(1 \leq i \leq N)} F(p),$$

we then obtain obviously

$$N \int F(f) W(f)\,df. \tag{20}$$

* By using the argument pointed out by Landau in § 4 of his paper "Theory of Superfluidity of Helium II" it may be noted that the existence of superfluidity is directly evident from the above mentioned properties of the function $E(f)$, following from the inequality: $v(0) > 0$. This inequality can thus be considered as the condition of superfluidity.

On the other hand, we have for the same mean value

$$\sum_f F(f)\bar{N}_f = \frac{V}{(2\pi\hbar)^3}\int F(f)\bar{N}_f\,df. \tag{21}$$

Thus the comparison of expressions (20), (21) gives

$$NW(f) = \frac{V}{(2\pi\hbar)^3}\bar{N}_f = N\frac{v}{(2\pi\hbar)^3}\overline{b_f^+ b_f},$$

whence we obtain, by expressing b_f, b_f^+ through ξ_f, ξ_f^+

$$W(f) = \frac{v}{(2\pi\hbar)^3}(1-|L_f|^2)^{-1}\overline{(\xi_f^+ + L_f^+\xi_{-f})(\xi_f + L_f\xi_f^+)} =$$

$$= \frac{v}{(2\pi\hbar)^3}\frac{\bar{n}_f + |L_f|^2(\bar{n}_{-f}+1)}{1-|L_f|^2}, \tag{22}$$

where, in virtue of (17)

$$\bar{n}_f = \left\{\exp\left(\frac{E(f)-(f\cdot u)}{\theta}\right)-1\right\}^{-1}. \tag{23}$$

Expression (22), obtained for the distribution function, holds evidently only for $f \neq 0$. Therefore, the complete expression for the momentum distribution function is seen to be, in virtue of the normalization condition (19):

$$W(f) = C\delta(f) + \frac{v}{(2\pi\hbar)^3}\frac{\bar{n}_f + |L_f|^2(n_{-f}+1)}{1-|L_f|^2}, \tag{24}$$

where $\delta(f)$ is Dirac's δ-function, and the number C is determined by the equation

$$C = 1 - \frac{v}{(2\pi\hbar)^3}\int \frac{\bar{n}_f + |L_f|^2(\bar{n}_{-f}+1)}{1-|L_f|^2}\,df. \tag{25}$$

C is obviously equal to N_0/N since CN represents the average number of molecules with zero momentum.

It follows from (11) that

$$\frac{|L_f|^2}{1-|L_f|^2} = \frac{\left(\frac{N_0}{V}\,\nu(f)\right)^2}{2E(f)\left\{E(f)+T(f)+\frac{N_0}{V}\,\nu(f)\right\}};$$

$$\frac{1}{1-|L_f|^2} = \frac{E(f)+T(f)+\frac{N_0}{V}\,\nu(f)}{2E(f)}. \tag{26}$$

Hence the momentum distribution function at absolute zero is of the form

$$W(f) = C\delta(f)+\frac{v}{(2\pi\hbar)^3}\,\frac{\left(\frac{N_0}{V}\,\nu(f)\right)^3}{2E(f)\left\{E(f)+T(f)+\frac{N_0}{V}\,\nu(f)\right\}},$$

whereby

$$1-C = \frac{v}{(2\pi\hbar)^3}\int\frac{\left(\frac{N_0}{V}\,\nu(f)\right)^2}{2E(f)\left\{2E(f)+T(f)+\frac{N_0}{V}\,\nu(f)\right\}}\,df. \tag{27}$$

We thus see that even at $\theta = 0$ only a fraction of molecules possesses momenta which are exactly zero, the rest being continuously distributed over the whole momentum spectrum.

As we have pointed out, our approximation method holds only when $(N-N_0)/N = 1-C \ll 1$, and, therefore, the interaction of molecules must be sufficiently small in order to secure the smallness of the integral (27).

We can now make clear the meaning of the assumed smallness of the interaction.

Let us put $\Phi(r) = \Phi_m F(r/r_0)$, where $F(\varrho)$ is a function assuming together with its derivatives the values of the order of unity for

$\varrho \sim 1$ and rapidly approaching zero as $\varrho \to \infty$. Then

$$\nu(f) = \Phi_m r_0^3 \omega\left(\frac{|f| r_0}{\hbar}\right),$$

$\omega(x)$ being a function assuming the values ~ 1 for $x \sim 1$, rapidly approaching zero as $x \to \infty$.

Transforming (27) to the dimensionless variables and reducing the three-dimensional integral to the one-dimensional one, we get

$$\frac{N-N_0}{N} = \frac{v}{r_0^3}\eta \frac{1}{(2\pi)^2}\int_0^\infty \frac{\eta\omega^2(x)x\,dx}{\alpha(x)\{x\alpha(x)+x^2+\eta\omega(x)\}}, \tag{29}$$

where

$$\alpha(x) = \sqrt{x^2+2\eta\omega(x)};$$

$$\eta = \frac{(r_0^3 N_0/V)\Phi_m}{\hbar^3/2mr_0^2} \sim \frac{(r_0^3/v)\Phi_m}{\hbar^2/2m_0^2}.$$

Now it is easy to see that for small values of η the integral in (28) is of the order of $\sqrt{\eta}$, and hence the condition of the validity of our method may be represented by the inequalities

$$\eta \ll 1; \quad \frac{v}{r_0^3}\eta^{3/2} \ll 1,$$

i.e.

$$\frac{r_0^2}{v}\Phi_m \ll \frac{\hbar^2}{2mr_0^2}; \quad \sqrt[3]{\frac{r_0^3}{v}}\,\Phi_m \ll \frac{\hbar^2}{2mr^2}. \tag{28}$$

For $\theta > 0$ an analogous consideration of the general formula (24) leads us to a supplementary condition of the weakness of the excitation which requires the temperature to be small as compared with the λ-point temperature.

It is to be pointed out that the inequalities (29) automatically exclude the possibility of accounting for the short-range repulsion forces, since this would require to admit the strong increment of $\Phi(r)$ in the vicinity of $r = 0$.

It seems nevertheless possible to modify the obtained results in the way to get them extended to the more real case of a low density gas of molecules possessing a finite radius.

To this aim we observe that the potential function $\Phi(r)$ appears in our final formula only in the form of the expression

$$\nu(f) = \int \Phi(|q|)e^{i\frac{(f\cdot q)}{\hbar}}\,dq, \tag{30}$$

proportional to the amplitude of Born's collision probability for binary collision.

Hence, as the interaction of molecules for the low density gas reveals itself principally by means of these binary collisions, it seems that expression (30) is to be replaced* by the corresponding expression proportional to the amplitude of the exact probability of the binary collisions, calculated for the limiting case of zero density, i.e. we have to put

$$\nu(f) = \int \Phi(|q|)\varphi(q, f)\,dq \tag{31}$$

where $\varphi(q, f)$ is the solution of the Schrödinger equation for the relative movement of an isolated pair of molecules:

$$-\frac{\hbar^2}{m}\nabla^2\varphi + \{\Phi(|q|) - E\}\varphi = 0,$$

going over into $e^{i\frac{(f\cdot q)}{\hbar}}$ at infinity. The replacement of (31) instead of (30) in the formula for $E(f)$ will lead us to the results, referring to low density gases.

This being admitted we see, e.g., that the condition of the superfluidity $\nu(0) > 0$ may be written in the form

$$\int \Phi(|q|)\varphi(|q|)\,dq > 0, \tag{32}$$

* I am indebted to L. D. Landau for this important remark.

where $\varphi(|q|)$ is the radially symmetric solution of the equation

$$-\frac{\hbar^3}{m}\nabla^2\varphi+\Phi(|q|)\varphi = 0$$

going over into unity at infinity.

In order to connect, as before, inequality (32) with the condition of the thermodynamic stability let us compute the principal term in the expansion of the free energy in powers of density at the absolute zero of temperature. The free energy at absolute zero being equal to the mean energy, we have the following expression for this energy per one molecule:

$$\mathcal{E} = \bar{T}+\frac{1}{2v}\int \Phi(|q|)G(|q|)\,dq, \tag{33}$$

where $\bar{T}$ is the mean kinetic energy of one molecule and $G(r)$ is the molecular distribution function normalized in the way that $G(r)\to 1$ as $r\to\infty$. On the other hand, by using the virial theorem we see that the pressure P can be determined by the formula

$$Pv = \frac{2}{3}\bar{T}-\frac{1}{6v}\int \Phi'(|q|)|q|\,G(|q|)\,dq. \tag{34}$$

Let us now remark that for $\theta = 0$ the principal term in the expansion of the molecular distribution function in powers of density is obviously equal to $\varphi^2(|q|)$. Therefore, by neglecting in (33), (34) the terms of the second order in density one gets

$$\mathcal{E} = \bar{T}+\frac{1}{2v}\int \Phi(|q|)\varphi^2(|q|)\,dq;$$

$$Pv = \frac{2}{3}\bar{T}-\frac{1}{6v}\int \Phi'(|q|)|q|\varphi^2(|q|)\,dq.$$

Hence, taking into account that

$$Pv = -v\frac{\partial\mathcal{E}}{\partial v},$$

we obtain the equation for the evaluation of the principal term in the expression for $\bar{T}$. In this way one gets

$$\mathcal{E} = \frac{1}{2v} \int \Phi(|q|)\varphi(|q|)\,dq = \frac{\nu(0)}{2v}, \quad P = \frac{\nu(0)}{2c^2},$$

and thus the condition of superfluidity in the considered case of low density gas is also equivalent to the usual condition of the thermodynamic stability at absolute zero:

$$\frac{\partial P}{\partial v} < 0.$$

It can also be seen that the energy of a quasi-particle goes over again into $c|f|$ for small f.

Consider now, for instance, the model of hard impenetrable spheres of the radius $r_0/2$ and put

$$\Phi(r) = +\infty, \quad r < r_0;$$
$$\Phi(r) = 0, \quad r < r_0.$$

Then by simple calculation we get

$$\nu(0) = 2\pi \frac{\hbar^2 r_0}{m}.$$

If a weak attraction between molecules is admitted here in the way that

$$\Phi(r) = +\infty, \quad r < r_0;$$
$$\Phi(r) = \varepsilon\Phi_0(r) < 0, \quad r > r_0.$$

ε being a small parameter, we obtain, up to the terms of the order of ε^2:

$$\nu(0) = 2\pi \frac{\hbar^2 r_0}{m} + 2\pi \int_{r_0}^{\infty} r^2 \Phi(r)\,dr.$$

Thus, the superfluidity in the considered gas model is conditioned by the play of repulsion and attraction forces, the first "encouraging" and the latter "hindering" it.

Let us note in conclusion that the extension of the present theory to the case of the real liquid seems possible if we be permitted to use such semi-phenomenological conceptions as that of the free energy of slightly non-equilibrium states.

T5. APPLICATION OF QUANTUM MECHANICS TO LIQUID HELIUM*

R. P. FEYNMAN

1. INTRODUCTION

Liquid helium exhibits quantum mechanical properties on a large scale in a manner somewhat differently than do other substances. No other substance remains liquid to a temperature low enough to exhibit the effects. These effects have long been a puzzle. It is supposed that they can all be ultimately understood in terms of the properties of Schrödinger's equation. We cannot expect a rigorous exposition of how these properties arise. That could only come from complete solutions of the Schrödinger equation for the 10^{22} atoms in a sample of liquid. For helium, as for any other substance today we must be satisfied with some approximate understanding of how, in principle, that equation could lead to solutions which indicate behavior similar to that observed.

Since the discovery of liquid helium considerable progress has been made in understanding its behavior from first principles. Some of the properties are more easily understood than others. The most difficult of these concern the resistance to flow above a critical velocity. If we permit some conjectures of Onsager[(1)], however, perhaps a start has been made in understanding even these. The aim of this article is to describe those physical ideas which have been suggested to explain the behavior of helium which can most easily be related to properties of the Schrödinger equation.

* *Prog. Low Temp. Phys.* **1**, 17 (1955).

We shall omit references to the phenomena involved in the Rollin film. It appears that the film can be understood as being maintained by van der Waals attraction to the wall. The flow properties of the film are interpreted as a special case of flow properties of helium in leaks in general.

The article falls naturally into two main sections. First, there are phenomena in which the superfluid velocity is irrotational. Here we can give a fairly complete picture. The second part concerns the case in which vorticity of the superfluid exists. Our position here is less satisfactory and more uncertain. It is described here in considerable detail because of the interesting problems it presents.

2. SUMMARY OF THE THEORETICAL VIEWPOINT

The first striking way that helium differs from other substances is that it is liquid even down to absolute zero. Classically at absolute zero all motion stops, but quantum mechanically this is not so. In fact the most mobile substance known is one at absolute zero, where on the older concepts we should expect hard crystals. Helium stays liquid, as London(2) has shown, because the inter-atomic forces are very weak and the quantum zero point motion is large enough, since the atomic mass is small, to keep it fluid even at absolute zero. In the other inert gases the mass is so much higher that the zero-point motion is insufficient to oppose the crystalizing effect of the attractive forces. In hydrogen the intermolecular forces are very much stronger, so it, too, is solid. In liquid ^{4}He there is a further transition at 2·2°K, the λ-transition, between two liquid states of different properties. A transition is expected (at 3·2°K) for such atoms if the interatomic forces are neglected, as Einstein(3) noticed. London(4) has argued that the λ-transition corresponds to the transition which occurs even in the ideal Einstein-Bose gas. The inter-atomic forces alter the temperature and, in a way as yet only imperfectly understood(5, 6) the order of the transition, but qualitatively the reason for the transition is understood. We will concern ourselves here, only with the liquid He II, below the λ-point, and shall try to elucidate the qualitative reasons for some of its strange behavior. Also we explicitly limit our considerations to a liquid made purely of ^{4}He atoms so that the wave function must be symmetric for interchange of the atoms. We do not mean to imply anything about liquid helium ^{3}He, nor about superconductors, either by analogy nor by contrast. That is, we shall use the fact that the wave function is symmetric in many arguments without stopping to inquire whether the symmetry is a necessary part of the argument.

The central feature which dominates the properties of helium II is the scarcity of available low energy excited states in the Bose liquid[(7, 8)]. There do exist excited states of compression (i.e.: phonons) but states involving stirring or other internal motions which do not change the density cannot be excited without expenditure of an appreciable excitation energy. This is because, for quantum energies to be low, long wave lengths or long distances are necessary. But the wave function cannot depend on large-scale modifications of the liquid's configuration. For a large-scale motion, or stirring, which does not alter the density, only moves some atoms away to replace them by others*. It is essentially equivalent to a permutation of one atom for another, and the wave function must remain unchanged by a permutation of atoms, because ^{4}He obeys the symmetrical statistics. The only wave functions available are those which change when atoms move in a way which is not reproducible by permutation, and therefore either, (1) movements accompanied by change in density (phonons), (2) movements over distances less than an atomic spacing, therefore of short wave

* In the ideal gas the low excitations are those in which one or two atoms are excited to low states. These involve density changes and are more analogous to phonon states (but are even lower in energy than in the liquid because the ideal gas has infinite compressibility, and therefore vanishing sound velocity). The interatomic forces in the liquid make it more imperative that if atoms are moved away from one point others move in to take their place, if high repulsive energies between nearby atoms are to be avoided.

length and high energy (rotons and more complex states), or (3) movements resulting in a change in the position of the containing walls (flow). We shall discuss these states in detail presently.

The scarcity of low energy excited states is the seat of many of the phenomena in the liquid. This has been known since the work of Landau who developed a theory of the liquid on the assumption of such scarcity. The specific heat is very low at low temperature and only rises rapidly above about 1°K when enough thermal energy is available to excite an appreciable number of the higher energy states (rotons). There are so few states excited that the excitations may be localized in the fluid like wave packets. These move about, collide with each other and the walls, and imitate the appearance that in the perfect background fluid there is another fluid or gas. This "gas" of excitations carries all the entropy of the liquid, may carry waves of number density (second sound, analogous to sound in an ordinary gas), finds it difficult to diffuse through long thin channels, tries to even up uneven velocity distributions among its roton "molecules" (viscosity), and acts in many ways as a normal fluid. Meanwhile the background in which the rotons travel, that is, the total body of fluid itself, can flow. It flows, at low velocity, without resistance through small cracks. The reason is that to have resistance, flow energy and momentum must go into heat, that is internal excitations (e.g. rotons). The energy required to form a roton is not available (at the necessary momentum change) unless the fluid velocity is very high.

Actually it appears likely that helium in flow doesn't form rotons directly at all. Resistance sets in at a relatively low velocity (critical velocity) because apparently a kind of turbulence begins in the perfect fluid*. This cannot occur at lower velocities because energy is needed to create vorticity. And, if we accept Onsager's suggestion, the vorticity is quantized, the line integral of the momentum per

* The author now considers his statement (reference 7) that the reason for flow resistance "cannot very well be a kind of turbulence", to be in error.

atom (mass of atom times fluid velocity) around a closed circuit must be a multiple of *h*. Below the critical velocity not enough kinetic energy is available in the fluid to produce the minimum vortex lines.

We shall discuss first the way that the scarcity of states accounts for many of the properties of the liquid. Here we are summarizing work of many others, particularly Landau. It is thought best to re-emphasize this viewpoint, since it is the one which is directly supported by quantum mechanics. Furthermore, in this way we are starting over the more familiar ground. Next we discuss the quantum mechanical view of the reason for the scarcity of states. Finally in the second part of the paper we discuss the quantized vortex lines proposed by Onsager.

3. LANDAU'S INTERPRETATION OF THE TWO-FLUID MODEL

One of the most fruitful ideas in interpreting the behavior of the liquid is the two-fluid model. It was developed by Tisza[(9)] from analogy to the structure of an ideal Bose gas. It is often spoken of as a vague association of two penetrating fluids. Landau[(10)] has interpreted it in a definite manner. We review his interpretation here, although an excellent review by Dingle[11] already exists. He has strongly emphasized the fact that one might picture the helium as a background fluid in which excitations move. At absolute zero one has a perfect ideal fluid which may flow frictionlessly with potential flow. If heated, the heat energy excites the liquid. This it does by creating here and there within it excitations of some sort. These excitations can make their way from one place to another, collide with the walls and with each other, and give to helium some properties associated with the so-called normal fluid component, such as viscosity. Landau, as a result of his study of quantum hydrodynamics, was lead to suppose the excitations to be of two kinds. Of lowest energy are the phonons, or quantized sound waves, whose energy E equals pc where p is the momentum and c the speed

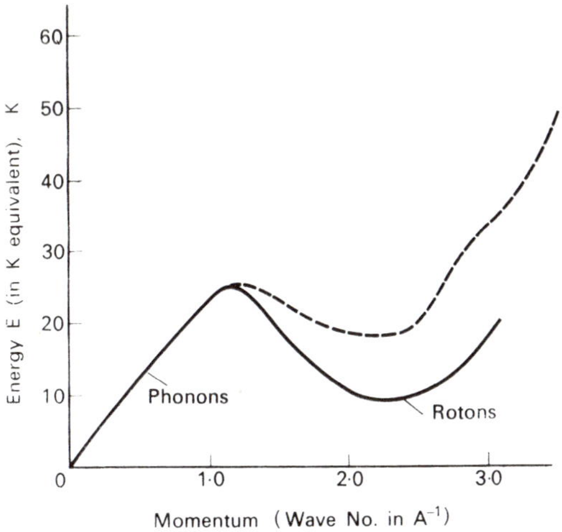

FIG. 1. The energy of excitations as a function of their momentum. Solid line as envisaged by Landau with parameters set to fit specific heat data; dotted line, as approximate curve derived from quantum mechanics. Excitations in linear section for low momentum correspond to phonons. Those near the minimum of the curve are called rotons.

of sound. Above these separated by an energy gap Δ are those of another kind, called rotons. At first he supposed the energy of these to be given by $\Delta+p^2/2\mu$ if they have momentum p, where μ is an effective mass. Later he found that this did not agree with the experiments of Peskhov on second sound, and he proposed instead the formula

$$E_{\text{rot}} = \Delta+(p-p_0)^2/2\mu \qquad (1)$$

where p_0 is some constant. He went further and suggested that all these excitations really are of the same class and differ only in momentum. The energy $E(p)$ of those of momentum p, depends only on the magnitude of p, rising at first linearly as pc, but later falling to a minimum at p_0 and rising again, as in Fig. 1, solid curve. The curve in the vicinity of the minimum is given by (1). At the low temperatures encountered in He II only the states near

$p = 0$, and those close to the minimum are excited. Therefore we do not have to know the rest of the curve accurately. Furthermore, the only important excitations are one of the two classes, phonons and rotons.

Supposing the excitations to obey Bose statistics the number, at temperature T, of momentum in the range $d^3\boldsymbol{p}$ is, according to statistical mechanics,

$$n_p = (\exp \beta E - 1)^{-1}\, d^3\boldsymbol{p}(2\pi\hbar)^3 \tag{2}$$

with $\beta = (kT)^{-1}$ and $E = E(p)$. From this the average energy $E(\boldsymbol{p})$ and the specific heat can be calculated. In agreement with experiment it begins at low temperature as T^{+3} as expected, according to Debye, since only phonons are excited. At higher temperatures the higher energy roton excitations become excited, and the specific heat rises much more rapidly. The thermodynamic properties are in excellent agreement with the theory if[12]

$$\begin{aligned} c &= 240 \text{ meters/sec} \\ \Delta/k &= 9{\cdot}6^\circ\text{K} \\ p_0/\hbar &= 2{\cdot}0\,\text{Å}^{-1} \\ \mu &= 0{\cdot}77m \end{aligned}$$

where m is the atomic mass of helium.

The hydrodynamic equations of the two-fluid model arise as follows. Suppose the fluid at absolute zero has density ϱ_o and velocity $\boldsymbol{v}_s$. In the first part of this paper we shall take $\boldsymbol{v}_s$ to be irrotational $\nabla \times \boldsymbol{v}_s = 0$. Later we discuss the problem of local circulation. The mass current density is $\varrho_o \boldsymbol{v}_s$ and the kinetic energy is $\frac{1}{2}\varrho_o \boldsymbol{v}_s^2$. Suppose that as a result of a rise in temperature a limited number of excitations are formed in the fluid. Landau has shown that the energy to form excitations in a moving fluid is not $E(\boldsymbol{p})$ but is

$$E = E(p) + \boldsymbol{p} \cdot \boldsymbol{v}_s \tag{3}$$

This results from simple considerations of the relations in moving and still frames of reference. The mass current density equals the momentum density of the fluid since all of the atoms have the same mass. It now is

$$\boldsymbol{j} = \varrho_o \boldsymbol{v}_s + <\boldsymbol{p}> \tag{4}$$

where $<\boldsymbol{p}>$ is the mean momentum of the excitations per unit volume. Now the mean $<\boldsymbol{p}>$ depends on how the excitations drift. If they are in equilibrium with the fixed walls of the vessel the mean $\boldsymbol{p}$ is *not* zero. The energy is $E(p)+\boldsymbol{p}\cdot\boldsymbol{v}_s$. It is lower than $E(p)$ for those excitations, whose momentum is directed oppositely to $\boldsymbol{v}_s$. Therefore in equilibrium more excitations align oppositely to $\boldsymbol{v}_s$ than parallel to it. For this reason the mean $\boldsymbol{p}$ is directed oppositely to $\boldsymbol{v}_s$ and for small $\boldsymbol{v}_s$ is proportional to it, let us say $<\boldsymbol{p}> = -\varrho_n \boldsymbol{v}_s$. This defines ϱ_n. If ϱ_s is defined as $\varrho_o - \varrho_n$ we have a total current $\varrho_s \boldsymbol{v}_s$ in a situation in which the excitations are in equilibrium with fixed walls. The equilibrium is established by collisions of the excitations with the walls and with each other.

The number of excitations of momentum $\boldsymbol{p}$ is again determined by (2) but now with E given by (3) so that the average $\boldsymbol{p}$ is

$$<\boldsymbol{p}> = \int \boldsymbol{p}\,(\exp \beta(E(\boldsymbol{p})+\boldsymbol{p}\cdot\boldsymbol{v}_s)-1)^{-1}\, d^3\boldsymbol{p}(2\pi\hbar)^{-3}$$

or expanding to first order in $\boldsymbol{v}_s$, find $<\boldsymbol{p}> = -\varrho_n \boldsymbol{v}_s$ where

$$\varrho_n = -\frac{\beta}{3}\int p^2\,(\exp(\beta E(p))-1)^{-2} \exp \beta E(p)\, d^3\boldsymbol{p}(2\pi\hbar)^{-3} \tag{5}$$

The density ϱ_n determined from experiments in second sound is in reasonable agreement with this expression (evaluated with the constants given above it fits above 1°K, but below 1°K the values $p_o/\hbar = 2{\cdot}3$ Å^{-1} and $\mu = 0{\cdot}40m$ fit better, and do not alter the good fit to the thermodynamic data). This explicitly shows that ϱ_n is a derived concept, and does not represent the density of anything which has microscopic meaning.

The excitations can drift also. The distribution for equilibrium in a drifting gas is, according to statistical mechanics,

$$n(E) = (\exp \beta(E-\boldsymbol{p}\cdot\boldsymbol{u})-1)^{-1} \tag{6}$$

where $\boldsymbol{u}$ is a parameter. In this case the mean momentum is

$$<\boldsymbol{p}> = -\varrho_n(\boldsymbol{v}_s - \boldsymbol{u}).$$

If we write $\boldsymbol{u} = \boldsymbol{v}_n$ we have for the current

$$\boldsymbol{j} = \varrho_0\boldsymbol{v}_s - \varrho_n(\boldsymbol{v}_s - \boldsymbol{u}) = \varrho_s\boldsymbol{v}_s + \varrho_n\boldsymbol{v}_n. \tag{7}$$

This can be interpreted macroscopically as saying that the current is like that in a mixture of two fluids, one of density ϱ_s moving at velocity $\boldsymbol{v}_s$, the other of density ϱ_n and velocity $\boldsymbol{v}_n$.

Actually (6) is not an equilibrium distribution unless the walls move at velocity $\boldsymbol{u}$, and furthermore $\boldsymbol{u}$ is constant throughout the liquid. It is generally taken as a good approximation in the case that $\boldsymbol{u}$, that is, $\boldsymbol{v}_n$, is not constant. The lack of equilibrium in this case produces irreversible effects, such as viscosity, which can be associated with the "normal fluid component". The distribution is in equilibrium even if $\boldsymbol{v}_s$ is not constant.

The entropy of the system is that of the excitations. It is easily verified that the mean group velocity of the excitations (the mean of $\delta(E(p)+\boldsymbol{p}\cdot\boldsymbol{v}_s)/\delta\boldsymbol{p}$) is just $\boldsymbol{v}_n$. The entropy can therefore be considered as flowing with the "normal fluid".

It is also possible to work out the expected value of the energy of the system. If one calculates the internal energy and subtracts the internal energy the system would have at the same entropy but with $\boldsymbol{v}_s = \boldsymbol{v}_n = 0$ the excess expanded to the second order in the velocities can be written $\frac{1}{2}\varrho_s v_s^2 + \frac{1}{2}\varrho_n v_n^2$. This is just what the two-fluid model would expect.

Therefore Landau shows that a liquid system with excitations as described will behave in many ways like a mixture of two fluids.

Furthermore, considerable progress has been made by Landau and Khalatnikov(13) in the interpretation of many irreversible phenomena, such as viscosity, attenuation of second sound, etc. from the kinetic theory implied by such excitations. It is not possible as yet to find the cross-section for collision, say between two rotons, from first principles. But if a few such quantities are considered as unknown parameters, a great deal can be said. The number of rotons varies very rapidly with temperature, in the manner given by (2). For this reason the mean free path for collision and the resultant viscosity resulting from roton-roton collisions has a known temperature dependence. In a similar way the contribution of collisions between rotons and phonons or between phonons can be worked out. There are also collisions in which the number of excitations changes. The results are often in excellent agreement with experiment.

There is, therefore, little doubt that in liquid helium there are such excitations, with the energy spectrum that Landau suggests, and that this picture supplies the complete interpretation of the two-fluid model for helium II.

4. THE REASON FOR THE SCARCITY OF LOW ENERGY STATES

The next question that concerns us is to try to see from first principles why the excitations of the helium fluid have these characteristics.

Landau has, in fact, tried to obtain some justification for the spectrum from a study of quantum hydrodynamics. This is not a completely detailed atomic approach. One attempts to describe the liquid by a few quantities such as density and current, or velocity. Then one makes these quantities operators with reasonable commutation relations, and tries to find the excitation energies of the fluid. The problem has not been analyzed in sufficient detail to establish the energy spectrum (1). Such an approach cannot give us an ultimate detailed understanding for two reasons. First, the numerical values of Δ, p_o, μ show these quantities to be characteristic of the atomic structure of the liquid. A theory which describes the fluid simply by average variables and which therefore cannot represent the fact that the liquid does in fact have atomic structure cannot lead to definite values for excitation energy. A more serious problem is this. It is necessary to show not only that the excitations

$E(p)$ exist, but that there are not a host of other possible excitations lying lower. If we describe the liquid with average variables we have no assurance that there are no excitations at a level below the coarseness of our averages. Possibly excitations exist which represent no gross density variation and no mean current. If many other lower excitations exist they dominate the specific heat curve and the properties of the fluid. (Perhaps in ^{3}He we have an example of a system capable of excitations at an atomic level which are not describable by the variables used in quantum hydrodynamics).

However it is possible from first principles to see why there are no other excitations but those supposed by Landau and why the energy spectrum of these excitations has, qualitatively, the form which he supposed.[7, 8]

In order to do so, we should, rigorously, have to solve the Schrödinger equation for the system

$$H\psi = -\frac{\hbar^2}{2m}\sum_i \nabla_i^2\psi + \sum_{ij} V(\boldsymbol{R}_{ij})\psi = E\psi$$

where m is the atomic mass, $V(\boldsymbol{R}_{ij})$ is the mutual potential of two atoms separated by the distance $\boldsymbol{R}_{ij} = \boldsymbol{R}_i - \boldsymbol{R}_j$, and ∇_i^2 represents the Laplacian with respect to the coordinates $\boldsymbol{R}_i$ of the ith atom. The sums must be taken over all of the N atoms in the liquid. We cannot solve this equation directly but we can make surprising headway in guessing the characteristics of the wave functions ψ which satisfy it.

We shall have to picture the wave function ψ. It differs from one state to another. But we will consider its value for only one state at a time. Then it is a definite but complicated function $\psi(\boldsymbol{R}_1, \boldsymbol{R}_2 ... \boldsymbol{R}_N)$ of the $3N$ variables $\boldsymbol{R}_i$. To picture it we must have a scheme by which we clearly represent it in our minds. Now such a function is a number associated with every set $\boldsymbol{R}^N$ of values $\boldsymbol{R}_i$, or, as we shall say, with every configuration of the atoms. We can represent a configuration $\boldsymbol{R}^N$ by imagining each of the N atoms in the vessel containing the liquid to be located with its center at one of the $\boldsymbol{R}_i$. That is, each configuration is represented, as classically, as a particular definite location for each of the atoms. Then $\psi(\boldsymbol{R}^N)$ is a number associated with each such arrangement of the atoms. We can call it the amplitude of the configuration. For a given state, this amplitude for some atomic arrangements is large—these arrangements then have large probability—for others small and the configuration is unlikely. When we wish to speak of how the amplitude changes as the values of $\boldsymbol{R}_i$ change, we shall use the more vivid language of asking how the amplitude changes as the atoms are "moved" about. Such motions are not directly related to any real classical motions, of course. In fact we cannot describe classical motions directly. All such classical ideas must be interpreted in terms of the mathematical behavior of ψ, if we are to be consistent with quantum mechanical principles. Most of our task, therefore, is trying to describe the ψ functions which correspond to the various kinds of states of energy, or motion, of which the liquid is capable.

Start by considering the ground state wave function which we shall call Φ. We use the intuition which we have acquired from knowing the solutions of the Schrödinger equations for simpler systems. For stationary states, ψ can be taken to be a real number. The lowest state always has no nodes (except for the exclusion principle, which does not operate here). Therefore Φ is everywhere positive. It is symmetrical, that is, Φ depends only on where atoms are, not on which is which. The energy $V(\boldsymbol{R})$ of interaction of two helium atoms, as worked out by Slater and Kirkwood,[14] for example, consists of a very weak attraction at large distances, but a powerful repulsion inside of 2·7 Å (see Fig. 2). The atoms in liquid helium at the normal density have a volume of 45 cubic Ångströms each so they are not tightly squeezed together. If one wishes a rough approximation, consider the atoms as impenetrable spheres of 2·7 Å diameter, and forget the attraction, whose effect is, after all, mainly just to hold the liquid together at the normal density even if the external pressure falls to zero. Then configurations of atoms in which some overlap each other, that is, are closer than about 2·7 Å, are of very small amplitude. In the most likely configurations the atoms are well spaced. As for a particle in a box whose wave function grows highest in the center and falls gradually to zero at the walls, we may imagine the amplitude highest for good separation and falling toward zero if a pair of atoms

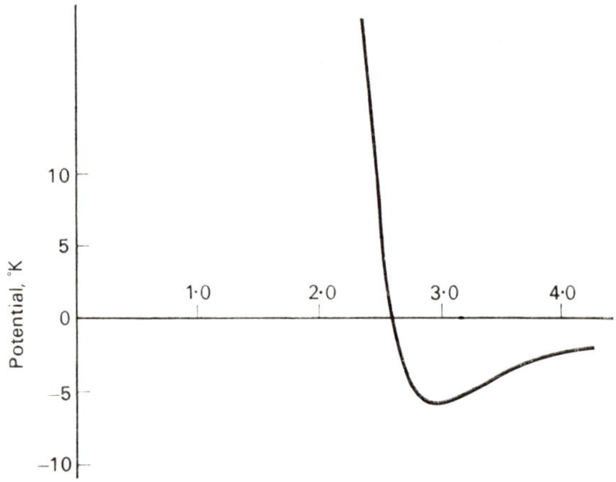

FIG. 2. The potential of interaction of two helium atoms as functions of their separation as worked out from quantum mechanics by Slater and Kirkwood

approach too closely. Our structure is a liquid, as a consequence of the zero-point energy, so that no particular lattice arrangement is strongly preferred. All configurations for which the spacing is ample have high probability. We can get from one arrangement to another without ever crossing a forbidden configuration of overlapping atoms because of the large spacing (cube root of atomic volume is 3·6 Å). Although not crystalline, there is a little local order induced by the tendency of atoms to stay apart, so that X-ray or neutron scattering experiments show a structure very similar to that of other simple liquids like liquid argon.

For the configurations of high amplitude the density is fairly uniform, at least until we look over such small volumes that we can see the fine grain atomic structure. If the density in a region is raised the atoms come closer together so that the "bow" on the wave function which occurs as one atom is moved from contact with a neighbor on one side to one of the other side, is confined to a smaller space. The increased curvature represents increased kinetic energy and it is not as likely to find a configuration in which such an energy barrier is penetrated. As a matter of fact, this feature is easily analyzed quantitatively. Long-range density fluctuations are sound waves. The rise of energy on compression is described by the compressibility coefficient, or equivalently by the speed of sound. Classically, standing density waves oscillating as a normal mode behave as an harmonic oscillator. Likewise, in quantum mechanics these are quantum oscillators and have zero-point motions, although the most likely configuration is that of uniform constant density. The wave function for the zero-point motion of an oscillator is a gaussian so that the amplitude Φ for a given kind of density fluctuation falls off exponentially with the square of the fluctuation. To summarize, the ground state function is large for any configuration in which the atoms are well spaced from one another at nearly constant average density. It falls off if these conditions are violated.

Next we turn to the excited states. Right away one obvious excitation is that of the standing sound wave. If the classical frequency is ω the quantum excitation energy of such a mode is $\hbar\omega$. Usually one prefers by linear combinations to make states of running waves, or phonons. If the wave number is k, the energy is $\hbar kc$ if c is the sound velocity.

We may readily obtain the wave function for such a phonon excitation. If the density is $\varrho(\boldsymbol{R})$ the classical normal coordinate going with such a mode is

$$q_k = \int \varrho(\boldsymbol{R}) \exp(i\boldsymbol{k}\cdot\boldsymbol{R})\, d^3\boldsymbol{R} \tag{8}$$

Quantum mechanically for an oscillator the wave function for the ground state is a gaussian, and the first excited state is just the coordinate times this gaussian (the first hermite polynomial $H_1(x)$ is just x). Hence the wave function is

$$\psi_{\text{phonon}} = q_k \Phi \tag{9}$$

if Φ is the ground state wave function of the system, which we have described in the preceding paragraphs. We have not bothered to normalize our function. The liquid consists of many atoms so if $\boldsymbol{R}_i$ is the position of the ith, the density in any configuration is

$$\varrho(\boldsymbol{R}) = \sum_i \delta(\boldsymbol{R}-\boldsymbol{R}_i) \tag{10}$$

the sum extending over all the atoms. Putting this in (8) and then (9) we find

$$\psi_{\text{phonon}} = (\sum_i \exp i\boldsymbol{k}\cdot\boldsymbol{R}_i)\Phi \tag{11}$$

This is valid if the wave length $(2\pi/k)$ is much larger than the atomic spacing, for then our description by compressional waves is adequate. The state energy is $\hbar kc$. Since $\hbar \boldsymbol{k}$ is the momentum $\boldsymbol{p}$ of the state, this means $E = pc$. Since the wave length can be very long this energy can be exceedingly low.

The central problem is to see why no states other than these phonons can have such low energies. We try to construct the wave function ψ of an excitation which should be as low in energy as possible and yet not represent a phonon. We must associate a number which may now be positive or negative with each configuration. In fact, since ψ must be orthogonal to the ground state Φ which is everywhere plus, ψ must be plus for half the configurations and minus for the other half. Furthermore, ψ must be orthogonal to all the phonon states. This simply means that ψ must vary from plus to minus for changes in the configurations which do not appreciably alter the large-scale density. Configurations can alter without variation of mean density by simply stirring the atoms about. Of course, since ψ must represent as low an energy as possible we must give low amplitude to configurations in which atoms seriously overlap, just as in the ground state Φ.

The function ψ takes on its maximum positive value for some configuration of the atoms. Let us call this configuration A, and the particular locations of the atoms α-positions. We said that the α-positions must be well spaced so that the atoms do not overlap, and further that they are, on a large scale, at roughly uniform

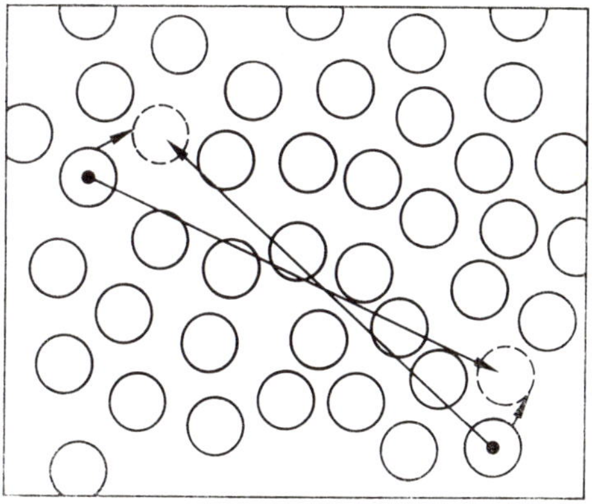

FIG. 3. Two configurations (solid and dotted) that result from large displacements (long arrows) of the atoms, can actually be accomplished by much smaller adjustments (short arrows) because of the identity of the atoms.

density. Equally, call configuration B, with atomic positions β, that for which ψ has its largest negative value. Now we want B to be as different as possible from A. We want it to require as much readjustment over as long distances as possible to change A to B. Otherwise ψ changes too rapidly and easily from plus to minus, our wave function has a high gradient, and the energy of the state is not as low as possible.

Try to arrange things so that *A requires* a large displacement to be turned into B. At first you might suppose it is easy. For example (see Fig. 3) in A take some atom in the left side of the box containing the liquid and move it way over to the other side of the vessel, and call the resulting configuration B. One objection to this is that an atom is moved from one side to another, so a hole remains at the left and an extra atom is at the right. This represents a density variation. To avoid this we may imagine that another atom has been moved at the same time from right to left, and the various holes and tight squeezes have been ironed out by some minor

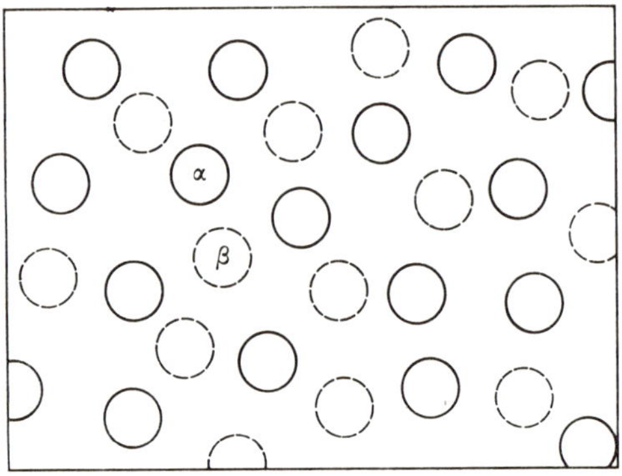

FIG. 4. The excited state wave function must be positive for one configuration, solid circles (α-positions) and negative for another. They are separated as far as possible if the negative configuration leaves no atom unmoved, dotted circles (β-positions)

adjustments of several of the neighbouring atoms. This movement of two atoms each a distance of the size of the vessel, one from left to right and the other from right to left, is certainly a long displacement, so B and A are very different. But they are not.

The atoms must be considered as identical, the amplitude must not depend on which atom is which. One cannot allow ψ to change if one simply permutes atoms. The long displacements can be accomplished in two steps. In the first step permute the atoms you wish to move to those α-positions closest to the ultimate position they are to occupy in the final configuration B. This step does not change ψ because all the atoms are still in the same configuration of α-positions. Then the change to the B configuration is made by small readjustments, no atom moving more than half the atomic separation. In this minor motion ψ must change quickly from plus to minus and the energy cannot be low. For the reason that the wave function is unchanged by permutation of the atoms it is impossible to get a B configuration very far from the A configuration. No very low energy excitations can appear (other than phonons) at all.*

In the phonon case we consider configurations in which, as ψ changes sign, the density distribution changes. A change in density cannot be accomplished by permuting atoms. That is why the Bose statistics does not affect phonon states. But it leaves them isolated as the lowest states of the system, so the specific heat approaches zero as T approaches zero according to Debye's T^3 law. This is the key argument for the understanding of the properties of liquid helium. It is given in somewhat more detail in reference 7. Since it is a negative argument, attempting to prove that a low energy state does *not* exist, it is difficult to convey conviction in a few words. The reader should try to invent wave functions of low energy for himself. After a few attempts he will see much more clearly what we have tried to explain here.

4. ROTONS

The qualitative argument is complete in itself. Nevertheless it is gratifying that it may be pushed even further to produce a quantitative estimate of the energy of these other excited states. We give only a summary of the considerations here (see reference 8 for details). We try to clarify our picture of the wave function ψ, until we can write a mathematical expression for it. This expression

* *(Note by Editor)*. The same conclusion follows from considerations by Bogolubov (1963) based on a Green functions approach and by Hohenberg and Martin (1965) based on a correlation functions approach.

put into the energy integral $\int \psi^* H\psi \, d^N V / \int \psi^* \psi \, d^N V$ will give us an estimate for the energy.

As we said, in order to get the energy as low as possible we wish the gradients of ψ to be small. Therefore the configuration B (where ψ is maximum negative) must be as far as possible from configuration A. Yet we noted that no β-site is more than half the atomic spacing from an α-site. The two configurations are generally nearly the same. They are furthest from each other if as many atoms as possible must be moved. That is accomplished when, as illustrated in Fig. 4, all the β-sites are between α-sites, so every atom must move. To completely specify ψ, of course, we must give its value for all configurations, not only for A and B when all atoms are on α-sites, or all on β-sites. The lowest energy results if the transition from plus to minus (hence A to B) is as gradual as possible. First for configurations in which each of the atoms is either on an α-or on a β-position, this is most naturally accomplished if ψ is proportional to the number on α-sites minus the number on β-sites. This difference passes smoothly from plus to minus. It can be expressed mathematically this way: Consider a function, $f(\boldsymbol{R})$, of position, which is $+1$ if $\boldsymbol{R}$ is at an α-site and -1 if $\boldsymbol{R}$ is at a β-site. Then $\sum_i f(\boldsymbol{R}_i)$ summed on all the atoms is just the desired number on α-sites minus number on β-sites. For intermediate positions ψ will vary as smoothly as possible if $f(\boldsymbol{R})$ is taken to vary in some smooth way between its extreme values of $+1$ and -1, which it takes on at α- and β-sites. This suggests that we take ψ to be of the form

$$\psi = \sum_i f(\boldsymbol{R}_i).$$

But this is incomplete for we tacitly assumed that in all the configurations the atoms did not overlap, the mean density did not vary very much and so on, just as in the ground state. This feature can be taken into account if we take instead

$$\psi_{\text{roton}} = \sum_i f(\boldsymbol{R}_i)\Phi \tag{12}$$

where Φ is the ground state function.[15] Then ψ will fall rapidly if the atoms overlap, etc. We actually do not know what the function $f(\boldsymbol{R})$ is but we expect it to vary rapidly, so that if expanded in a fourier integral the dominant wave lengths would be the atomic spacing.

According to the variational principle the best wave function is that which minimizes the energy integral. In this way, by variation of $f(\boldsymbol{R})$ it is readily found (see reference 8) that the minimum results if the function is

$$f(\boldsymbol{R}) = \exp(i\boldsymbol{k}\cdot\boldsymbol{R}) \tag{13}$$

and that the corresponding energy is*

$$E(k) = \hbar^2k^2/2mS(k) \tag{14}$$

where m is the atomic mass. The function $S(k)$ is the form factor for the scattering of neutrons from the liquid. That is, it is the Fourier transform of the function $p(\boldsymbol{R})$ which gives the probability per unit volume of finding an atom at a distance $\boldsymbol{R}$ from a given atom in the liquid in the ground state.

The local partial order of the liquid in the ground state shows up as in other liquids as a ring in the diffraction pattern (of neutrons, or X-rays). That is to say, there is a maximum in the function $S(k)$, which occurs when k represents a wave length near the nearest neighbor spacing. The maximum in $S(k)$ represents a minimum in $E(k)$ here. This confirms the expectation that the low excitation would have wave numbers in this vicinity.

The state (12) and (13) has the momentum $\boldsymbol{p} = \hbar\boldsymbol{k}$. Ordinarily not every value of a parameter in a wave function has significance in the variational method. But states of different momenta are orthogonal, and the energies (14) are significant not only for k near the minimum, but also in the neighborhood of this value. The range of values for which (14) is useful is limited only by the range for which (12) can be expected to be a good wave function. For

* *(Note by Editor.)* For a derivation of equation (14) see the paper by Pitaevskii reprinted as paper T7 in this volume (p. 335).

small k, (12) is identical to the wave function (11) representing phonon excitation, and (14) can be shown to give $\hbar kc$ in that region. Therefore the expression should be reasonable not only for k near the reciprocal atomic spacing, but for low k as well. It predicts a spectrum at first linear in $p(=\hbar k)$ then falling to a minimum, just as anticipated by Landau, and in agreement with experiment.

The curve $S(k)$ taken from neutron data of Henshaw and Hurst(16), or from the X-ray scattering data of Reekie(17) agree. The $E(k)$ which results is shown in Fig. 1 by the dashed line. The general behavior and minimum are clearly shown.

The actual value of the energy at the minimum is twice too high to agree with the experimental value (solid line) for Δ. The theoretical value lies above the true value, as it should according to the variational principle.

The inaccuracy of the wave function (12) prevents us from giving a complete description of what the roton wave function must look like. The function (12) does not satisfy the conservation of current. It appears as though a more accurate function would represent a current distribution large and unidirectional in one region, with a field of return currents surrounding it, somewhat in the nature of a smoke ring. These and other arguments suggest a trial function of the type*

$$\psi = \sum_i \exp i\boldsymbol{k}\cdot\boldsymbol{R}_i \cdot \exp i \sum_j g(\boldsymbol{R}_j - \boldsymbol{R}_i)\Phi \tag{15}$$

with the g, representing the back flow, to be determined. It is very hard to perform the integrals required in the variation problem with (15), so it has not been verified whether (15) represents a substantial improvement.

One way to understand the low energy for k near the reciprocal atomic spacing is this. One might consider these as sound waves of very short wave length. To obtain a density variation of long wave length is hard. To make the compression work must be done against

* *(Note by Editor.)* Compare Feynman and Cohen (1956).

opposing forces. For wave lengths closer to atomic spacing, however, such density variations are easier to arrange. In fact, one can create variations of wave length equal to the atomic spacing simply by arranging the atoms, doing no appreciable work against repulsions, the energy being purely kinetic $\hbar^2k^2/2m$. Actually the energy is even lower ($S(k)$ at maximum is 1.3) for there is a positive tendency in the liquid to have such variations; if some atoms are correctly arranged the others are more likely to be also satisfactory because of the local order. Therefore the energy does not continue to rise as $\hbar kc$ but falls lower for wave lengths near the atomic separation.

It is easy to verify that these excitations behave in just the way that has been assumed in developing the statistical mechanics and the two-fluid model. To represent a state with two excitations, say with momenta $\boldsymbol{k}_1$ and $\boldsymbol{k}_2$, one has the approximate wave function

$$\psi = (\textstyle\sum_i \exp i\boldsymbol{k}_1\cdot\boldsymbol{R}_i)(\sum_j \exp (i\boldsymbol{k}_2\cdot\boldsymbol{R}_j))\Phi$$

and so on. Since the order of the factors is irrelevant this is the same state if $\boldsymbol{k}_1$ and $\boldsymbol{k}_2$ are reversed. The excitations obey the Bose statistics. In a moving fluid the energy of the excitations can be shown to be (3).

6. IRROTATIONAL SUPERFLUID FLOW

So far we have only described the wave function for states representing internal excitation. We turn next to a description of the wave function which represents the state of the fluid when macroscopically we say it is flowing. We will assume that the flow velocity does not vary appreciably over distances of the order of an atomic spacing.

It is not difficult to represent by wave functions states which represent the motion of the superfluid. Suppose the system is at absolute zero so there are no excitations. If the entire system moves forward as a body, since the center of gravity coordinate can be separated out from Schrödinger's equation, the wave function is

$$\psi = (\exp i\boldsymbol{k}\cdot(\textstyle\sum_i \boldsymbol{R}_i))\Phi$$

where $N\hbar\boldsymbol{k}$ is the momentum of the system, if there are N atoms. In case the velocity is not uniform we can construct a wave function somewhat as follows: If the velocity varies only slowly from place to place, those atoms temporarily in a macroscopic region where the velocity is, say, $\boldsymbol{v}$ must surely have a wave function very much the same as though the liquid in the region were isolated and moving at a uniform velocity. This suggests that the phase contains a term $\hbar^{-1}m\boldsymbol{v}\cdot\sum_i \boldsymbol{R}_i$, the sum being taken only over those atoms in the region. Other regions where $\boldsymbol{v}$ differs make similar contributions to the phase so the total phase is $m\sum_i \boldsymbol{v}(\boldsymbol{R}_i)\cdot\boldsymbol{R}_i$ where $\boldsymbol{v}(\boldsymbol{R})$ is the velocity at $\boldsymbol{R}$. This suggests a wave function of the form

$$\psi_{\text{flow}} = [\exp i(\textstyle\sum_i s(\boldsymbol{R}_i))]\Phi \tag{16}$$

where $s(\boldsymbol{R})$ is a function which varies only very little over distances as small as the atomic spacing. We have suggested that it is $\hbar^{-1} m\boldsymbol{v}(\boldsymbol{R})\cdot\boldsymbol{R}$, but as is usual for waves whose wave length varies with position, the momentum is the gradient of the phase, not the coefficient of $\boldsymbol{R}$. Thus (16) does represent the helium flowing, but the velocity is given by

$$v = \hbar m^{-1}\nabla s. \tag{17}$$

It is readily verified that the current density is $\varrho_o \boldsymbol{v}$, and the energy (from the variational integral) is $\frac{1}{2}\varrho_i v^2$, as expected classically. There is no change in density, as in (16) we have not allowed these small effects to be represented.

If excitations exist in the moving fluid the wave function is (16) multiplied by the factor $\sum_i f(\boldsymbol{R}_i)$ in (12). The excitation energy turns out to be (3) as expected, interpreting $\boldsymbol{v}$ as the superfluid velocity $\boldsymbol{v}_s$.

Equation (17) implies that the motion is irrotational, that is, $\nabla\times\boldsymbol{v}_s = 0$. In a simply connected region this has only one solution for given motion of the boundaries. For fixed boundaries it is $v_s = 0$. In a multiply connected region the situation is different. Since $\nabla \times v_s = 0$, the circulation about any closed curve which can be shrunk to a point is zero. On the other hand, in the case of a toroidal region, if the curve encloses the hole the circulation need not vanish. Although the wave function must be single valued, s may be of the nature of the azimuthal angle, increasing by 2π, or a multiple thereof if one goes around the hole. That is, for a circuit enclosing a hole (into which liquid may not freely flow) the circulation must be an integral multiple n of a quantized unit $2\pi\hbar/m$,

$$\oint v_s \cdot ds = 2\pi\hbar m^{-1} \cdot n = 2\pi n \cdot 1{\cdot}5 \times 10^{-4}\ \text{cm}^2/\text{sec}. \tag{18}$$

These states do not influence the previous statistical mechanical argument. There are too few of them. The velocity may be considered as a macroscopic variable, such as density. For a macroscopic torus even the lowest of the states given by (18) is very much higher than a roton energy Δ. Thus if the torus area is A, radius R, the mass moving is $mA \cdot (2\pi R)/d^3$ where d^3 is the atomic volume. It moves at velocity given by $v_s \cdot 2\pi R = 2\pi\hbar m^{-1}$, from (18), so the kinetic energy is $(\hbar^2/2md^2) \cdot (2\pi A/Rd)$. The factor $\hbar^2/2md^2$ is an energy of the order of a roton, but the second factor is very large, being the torus dimension over the atomic spacing. Incidentally the total angular momentum is $\hbar$ per atom.

If the fluid must flow irrotationally, at first sight, it cannot lose energy, unless it is moving very rapidly. This has been pointed out by Landau. If a body of fluid is moving at velocity v, and loses a small energy δE, it must do so (to keep the flow irrotational) by the entire fluid changing its velocity. Let the change in v be δv. If M is the effective fluid mass the momentum change δp is $M\delta v$ and $\delta E = Mv\delta v = v\delta p$. Now this energy loss must go into heat; that is, into internal excitations of rotons. But if the momentum transferred to excitations is δp the energy cannot be small. It must be at least about $(\delta p/p_0)\Delta$ where Δ and p_0 are the energy and momentum of an individual roton. That is, δE must be at least $(\Delta/p_0)\delta p$ and energy cannot be lost unless v exceeds Δ/p_0, about 70 meters per second. (More accurately v must be high enough that a line drawn from the origin at slope v can cut the $E(p)$ *vs* p curve). This suggests the reason for the frictionless flow of superfluid. But we have proved too much, for in actuality the resistance sets in at velocities a few hundred times smaller.

The only way that gross slowing down can occur for lower velocities is for small parts of the fluid to stop or slow down without the entire fluid having to slow down at once. That is, energy loss must be accompanied by flow which is not irrotational; that is, flow which involves local circulation. To understand such effects we must add a new element to our picture of phonons, rotons and potential flow. These are the quantized vortex lines suggested by Onsager.[1] We proceed to describe them.

7. ROTATION OF THE SUPERFLUID

The problem which now faces us is to extend (16) so that we can also represent states for which $\nabla \times \boldsymbol{v}$ does not vanish, or at least where there is circulation in the superfluid. We analyze the situation at absolute zero for simplicity. We must present ourselves a problem in which such circulation is necessary and try to find the lowest energy state. The situation first considered by the author was the slip-stream between two regions of fluid moving at different velocity, but it is easier to arrive at the result by considering the problem of helium with high angular momentum in a cylindrical vessel. Suppose, for example, the helium at absolute zero is initially under such pressure that it is solid and is set into rotation, then the pressure is released so that it liquefies. What is the final state of the helium? We ask then for the lowest state of a quantity of helium which has a definite, macroscopically high, total angular momentum.

For a system of given angular momentum the kinetic energy is least if the angular velocity ω is a constant throughout the liquid. This motion is not rotation free for $\nabla \times \boldsymbol{v} = 2\omega$. But it is very difficult for helium to manage a state of local circulation. In fact, without high excitation energy, local circulation is impossible. At first one might find it hard to see why the liquid cannot simply rotate as a rigid body. The energy is then low. But a liquid is not a rigid body. A part of it can turn independently of the whole. In a rough way of speaking the liquid may be thought of as made up of many quasi-independent units of nearly atomic dimensions. Any motion of the body can be compounded of motions of the tiny parts. But to set any small part into a rotational state requires a high energy because the moment of inertia is so small. If only a limited energy is available nearly all the "parts" must be frozen out in their ground states. That is, nearly everywhere the local angular momentum is zero, i.e., $\nabla \times \boldsymbol{v}_s = 0$. It takes energy to create circulation and, furthermore, we can expect this circulation not to be distributed uniformly throughout the fluid. The rigid body type of rotation where $\nabla \times \boldsymbol{v}_s \neq 0$ everywhere is not possible, or if at all, only with an enormous expenditure of energy, an expenditure far higher than that gained by the uniform distribution of angular velocity.

Another possibility that suggests itself is that the liquid, if the angular momentum is high, is not free of excitations like rotons and phonons even though the temperature is at absolute zero. These excitations could carry the angular momentum. That is, in the language of the two-fluid model, perhaps there is at $T = 0$ a mixture of superfluid and normal fluid, with the superfluid component not rotating, and with the normal fluid carrying all of the angular momentum. The energy to maintain the normal fluid being sustained by the fact that if less normal fluid were present, for given angular momentum the kinetic energy would have to be larger. This turns out, for vessels of centimeter dimensions turning at about one radian per second, to be a state of nearly 10^4 times the energy of a rigid body rotating at the same angular velocity. Surely nature can find some lower state for the helium.

We know (see 18) that if there is a hole in the liquid, circulation can exist. Therefore another solution suggests itself. The liquid circulates around a hole with constant circulation as in a free vortex (familiar from rotation of water around an emptying drain). The velocity varies inversely as the radius, rising to such heights near the center as to be able to maintain the hole free of liquid by centrifugal force. Such a solution would be easy to verify in a striking manner by looking at the surface of the liquid. Instead of the usual parabola it would be the curve of the surface of a free vortex. The energy is still quite a bit higher than the rigid body case, because the velocity, instead of being distributed proportionally to the radius, actually falls as the radius increases. Nevertheless it is orders of magnitude below the mixture of normal fluid suggested above.

However, this is still not the lowest possible energy state, and the striking experiment will not succeed. To show this we construct a lower state. Suppose that the liquid has not only one vortex at the center, but several vortices. For example, suppose beside the central one there were a number distributed about the circle of radius $R/2$, half that of the vessel R, and all turning the same way. Viewed grossly this is like a vortex sheet so the tangential velocity can jump as we pass from inside $R/2$ to outside. Then the velocity can be arranged a little more like the linear curve by two sections, each of which is a $1/r$ curve. The gain in energy resulting from this improved distribution may more than compensate the energy needed to make the additional holes (and, further, the central vortex need not now be so large and energetic).

Continuing in this way with ever more vortices it soon becomes apparent that the energy can always be reduced if more vortices form. However, there is a limit. Due to the quantization (18) of the vortex strength the smallest vortex has circulation $2\pi\hbar m^{-1}$. The lowest energy results if a large number of minimum strength vortex lines (which we shall call unit lines) form throughout the fluid at nearly uniform density. The lines are all parallel to the axis of rotation. Since the curl of the velocity is the circulation per unit area, and the curl is 2ω, there will be

$$2m\omega/2\pi\hbar = 2{\cdot}1\times 10^{3}\omega \text{ lines per cm}^{2} \tag{19}$$

with ω in radians per second. For $\omega = 1$ rad per second the lines are about 0·2 mm apart so that the velocity distribution is practically uniform.

Such weak lines will not form actual macroscopic holes. In fact, if one neglects atomic structure and assumes a classical continuous liquid with surface tension, a unit line makes a hole opposed by surface tension which figures out to be only 0·4 Å in radians. That means that there is no real hole in the liquid. Around such a unit line, for example a straight one along the z-axis, the wave function off the axis is roughly

$$\psi = (\exp i \textstyle\sum_i \varphi_i)\Phi \tag{20}$$

where φ_i is the angle about the z-axis. This does not hold close to the axis. On the axis $\exp i\varphi$ is meaningless, and close to it has enormous gradients. A particle on the axis cannot have angular momentum, yet (20) implies that each atom has angular momentum $\hbar$, nor can there be exceptions because the Bose statistics implies that they are equivalent. Therefore a more accurate expression than (20) would be this expression multiplied by a factor which is unity except if any one of the atoms comes very close to the axis, in which case it falls rapidly to zero. The density of fluid falls to zero on the axis. This is the remnant of the classical hole. Actually quantum mechanically the line will not remain perfectly straight in one spot but will have some zero point motion of wandering and waving to and fro.

It is not hard to get a reasonable estimate of the energy contained in these lines. First consider an isolated unit line along the axis of a cylinder of length L, radius b. The velocity at radius r is $\hbar/mr$ and if

ϱ_0 is the fluid density in atoms per cc ($\varrho_0 = 1/45$ Å^3) the kinetic energy is the integral

$$K.\,E. = \tfrac{1}{2}\int \varrho_0 m(\hbar/mr)^2 \cdot 2\pi\, dr \cdot L.$$

The upper limit of the integral is b. It diverges at the lower limit, but within about the atom spacing the velocity formula is meaningless. Furthermore, inside this radius some of the energy is potential, required to keep to density down near the axis (that is, to make the partial "hole"). Therefore the energy needed to form such a line, per unit length, is

$$\begin{aligned}\text{Line energy per unit length} &= \varrho_o \pi \hbar^2 m^{-1} \ln(b/a) \\ &= 10^{-8} \ln(b/a) \text{ ergs/cm.} \qquad (21)\end{aligned}$$

Here a is a length of order of the atomic spacing. Its exact determination would require solving the difficult quantum mechanical problem. In almost all application the ratio b/a will be very large, and the logarithm large enough to be insensitive to the exact value of a. For this reason we will not attempt a detailed evaluation, but simply choose a to be close to the atomic spacing. We arbitrarily take $a = 4{\cdot}0$ Å. In more complicated geometrical situations the lower limit will be the same, but the upper limit b will be some other characteristic dimension of the apparatus, or more usually the spacing between vortex lines, etc. It can be found by integrating the velocity distribution as determined for the given distribution of singular vortex lines.

For a cylinder of liquid rotating at angular velocity $\omega = 1$ rad/sec the vortices are about 0·02 cm apart. This is $0{\cdot}5 \times 10^6$ times a if $a = 4$ Å, so we can take the $\ln(b/a)$ in this case to be about $\ln(0{\cdot}5 \times 10^6)$ or 14. Neglecting the variation of this logarithm with ω we find for the energy of all of the lines:

$$\text{Total line energy per unit volume} = \varrho_o \omega \hbar \cdot \ln(b/a)$$

where we have estimated $\ln(b/a)$ as 14. The ratio of this to the kinetic energy for a rigid body is $4\hbar m^{-1} R^{-2} \omega^{-1} \ln(b/a)$ if the cylinder radius is R. For $R = 1$ cm, $\omega = 1$ rad/sec this ratio is 10^{-2}.

For macroscopic laboratory dimensions the excess energy to form the lines is small. They would form if rotating solid helium is melted by releasing the pressure, the angular velocity distribution would differ imperceptibly from uniformity, and the surface should appear parabolic.

It is not self-evident that there is no state of appreciably lower energy, and that the energy of the rotating liquid is correctly estimated. This subject has not yet been analyzed any more deeply than is reported here. Therefore this part of the paper is not on as firm a foundation as the rest. We must therefore still consider it conjectural whether the considerations on rotational flow reported here are actually correct. It is interesting that all the conclusions were arrived at independently by the author without knowledge of Onsager's previous work (with which they are in exact concordance).

8. PROPERTIES OF VORTEX LINES

In a situation more general than uniform rotation, in which the curl of the velocity is not constant, we can imagine a similar situation. We have a situation instantaneously with many vortex lines. Some are closed on themselves in rings, and others terminate with their ends on the fluid boundaries. Viewed from a continuum approximation in which atomic structure is neglected, a velocity $\boldsymbol{v}_s$ can be defined at every point. The curl of this is zero everywhere, except at one of the vortex lines where it is infinite. These lines are real quantized vortex lines. The circulation around a small circuit surrounding only one line is $2\pi\hbar m^{-1}$. The lines have a sense depending on the direction of rotation. The circulation about any curve whatsoever is given by

$$\oint \boldsymbol{v}\cdot\boldsymbol{d}_s = 2\pi\hbar m^{-1}n$$

where n is always an integer, being the net number of lines linked by the circuit, account being taken to the sign of each.

If $\nabla \times \boldsymbol{v}_s$ is averaged over a large enough region that many lines are included, the number of lines per cm^2 must be at least $<\nabla \times \boldsymbol{v}_s> m/2\pi\hbar$ and the energy of these lines per unit volume is at least

$$\tfrac{1}{2}\,|<\nabla \times \boldsymbol{v}_s>|\,\varrho_0\hbar\cdot\ln(b/a) \tag{22}$$

where b is the spacing between lines, $1/b^2 = <\nabla \times \boldsymbol{v}_s> m/2\pi\hbar$. This shows that in our liquid it takes energy to create circulation. Actually in real, complex situations the energy might exceed greatly the value in (22). There may be great complex activity with many lines twisting and turning so that several lines of opposite senses are close together. In this case, the case of developed turbulence, the number of lines present may be bigger than the average $\nabla \times \boldsymbol{v}_s$ would indicate. (Probably in such a case it would be hard to define the average $\nabla \times \boldsymbol{v}_s$ because the result may depend on the size of the region over which the average is taken).

The discussion of the rotating cylinder of liquid with which we introduced the lines is rather special. We shall try to give a more complete and general description of the state of the superfluid with circulation. We continue to study the case at absolute zero. Let us try to characterize the state of a fluid in which we desire two things (which, it will turn out, are mutually incompatible). We want (a) the liquid to be flowing with a velocity $\boldsymbol{v}_s$ which is a smooth function of position without singularities (on a scale of distances large compared to atomic dimensions) and (b) we want $\nabla \times \boldsymbol{v}_s$ not to vanish.

Suppose the liquid in an element of volume ΔV (large compared to the atomic volume) is moving at velocity $\boldsymbol{v}$. Then as we have seen the wave function should depend on the position of the atoms, if they are within ΔV as exp $(im\boldsymbol{v}\cdot\sum_i \boldsymbol{R}_i)\,\boldsymbol{\Phi}$. That is, if a number of atoms in the region are displaced, each by $\Delta \boldsymbol{R}_i$ from one allowed (by $\boldsymbol{\Phi}$) configuration to another allowed one, the main effect is that the wave function must change phase by

$$\textstyle\sum_i (m\boldsymbol{v}\cdot\Delta \boldsymbol{R}_i)\hbar^{-1} \tag{23}$$

This can also be seen in another way. If a region of fluid can be considered to have a velocity $\boldsymbol{v}$ it has a momentum density $\varrho_0 m\boldsymbol{v}$. It is characteristic of momentum in quantum mechanics, that if the center of mass is changed the wave function changes phase by an amount proportional to the momentum and to the displacement of the center of mass. Now if the atoms are displaced by $\Delta\boldsymbol{R}_i$ the center of mass moves so the phase change (23) results. This is true at least if the displacement makes no other important change in the wave function. We will suppose that both before and after the displacement the atoms are well spaced and there are no gross density fluctuations, etc. such that in case the liquid were not in motion both configurations would have essentially the same amplitude.

The same argument goes for atoms in other regions, etc. so the phase shifts accumulate to a sum in (23) over displacements of atoms all over the liquid, if $\boldsymbol{v}$ is now considered as a function of $\boldsymbol{R}_i$. The displacements $\Delta\boldsymbol{R}$ must be small compared to the distances over which $\boldsymbol{v}$ varies. We shall apply the formula in a case in which $\Delta\boldsymbol{R}$ is the separation between atoms.

Select, in a given configuration, a very long closed chain of atoms each of which is a nearest neighbor of the next in line (see Fig. 5). The last should have the first as nearest neighbor. The chain may consist of very large numbers of atoms and may even be so long that it passes through regions of varying velocity. Consider a displacement of each atom to its nearest neighbor next in line. The wave function cannot change, for it is simply a special permutation of the atoms. Further we will suppose that if all the displacements are made together a little at a time, each intermediate configuration is allowed. This sliding of the chain along itself is not prevented by potential barriers, especially if we allow small temporary displacement of other atoms adjacent to the ring to permit passage in tight places. In the final configuration all atoms have returned to their original positions, except those of the ring which have moved one over. We suppose, because of the ease in which the displacement can be made that we can assume the wave function does not vanish for any intermediate position during the displacement. Then its phase shift is given by (23), but this must represent no change in the wave function. It is therefore necessarily an integral multiple of 2π. We conclude that

$$\oint \boldsymbol{v}_s \cdot d\boldsymbol{s} = 2\pi\hbar m^{-1} n \tag{24}$$

where n is an integer and the integral is taken over any path which

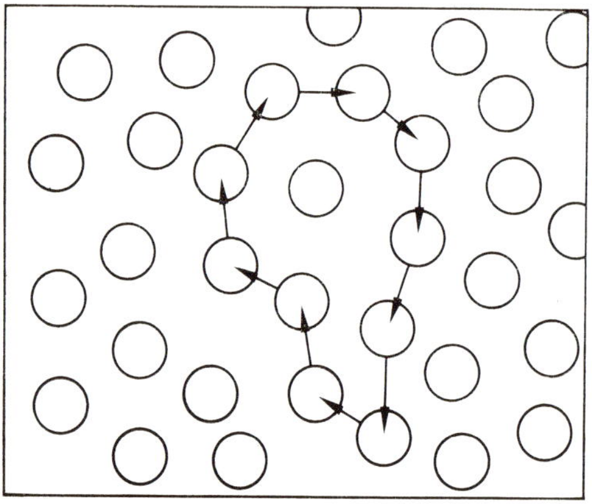

FIG. 5. The wave function must not change as a result of a permutation. If all the atoms are displaced around a ring, as shown, the phase change must be a multiple of 2π.

goes from one atom to the next neighbor, etc. If $\boldsymbol{v}_s$ is now assumed continuous at an atomic scale, the path can be smoothed out to any continuous curve. Of course, it is impossible that (24) holds for all continuous paths if n is an integer (depending on the path) if $\boldsymbol{v}_s$ is free of singularities and continuous unless $n = 0$ (in a simply connected region). Because any path can be deformed continuously into an infinitesimal path, the left side changing continuously to zero. The right side cannot change continuously so it must be zero for all paths. Likewise for a toroidal region n must be the same for all paths which surround the hole.

We see therefore that $\boldsymbol{v}_s$ cannot be continuous if we are to have circulation. There must be places where $\boldsymbol{v}_s$ is discontinuous, and places in the fluid where a displacement of an atom to its neighbor may not be possible without passing through a node in wave function. In the neighborhood of such a node the probability of finding an atom is reduced. This decrease in density requires energy to maintain it. We shall therefore try to arrange conditions so that such places are as infrequent as possible. Under those conditions, for nearly every conceivable ring of atoms the atoms can be moved over to the next adjacent atom without the wave function vanishing. Its phase change must be a multiple of 2π. If two adjacent rings have a phase change which is different, differing by 2π say, then between them somewhere must lie a very small ring of three or four atoms for which the circulation is $2\pi\hbar/m$. For example, suppose for a certain ring A the phase is zero, but for a nearby ring B it is 2π. Then shift ring B by a few atoms at a time until it gets as close to ring A as possible, but still has phase shift 2π. Likewise shift A until it is as close to B as possible but so that it has still shift 0. Then A and B will contain many atoms in common and only differ by a few, as illustrated in Fig. 6. Then consider a permutation consisting of shifting B forward, then shifting A backward. It is readily verified that this change is the same as a shift of atoms around the very small ring C consisting of those parts of A and B which are not common, plus one of the common atoms. But the change in phase is 2π when B shifts and 0 when A shifts back, so that it must be 2π for the very small ring C.* This represents a highly concentrated angular momentum. Somewhere in the middle of ring C is a nodal point. It is readily appreciated geometrically that these nodal points must essentially

* The change in phase cannot be determined only from the initial and final configuration, but requires a description of the amplitude for intermediate configurations as well. Therefore this argument is not complete unless it is also assumed that partial rotations of C consisting of displacements of less than one atom spacing can also be roughly imitated by partial displacements of B and A.

form lines through the fluid. They are quantized vortex lines. It must be admitted that this argument is far from complete. We should consider states in which the location of the vortex line is uncertain, that is, a superposition of states with various locations for the line. Such a state would have a lower energy. Possibly we make a serious error in imagining that the velocity can be defined right up to atomic distance from the axes, if this axis itself does not have a definite location. Onsager has remarked, in private

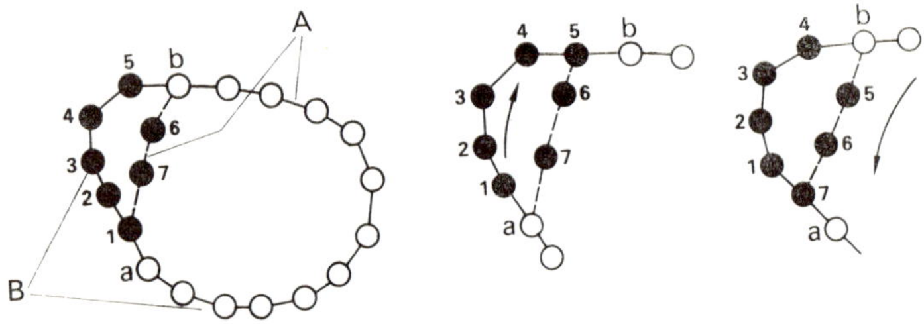

FIG. 6. A displacement along ring *B* followed by a reverse displacement along an adjacent ring *A* with many atoms in common is equivalent to a displacement around the ring *C*, indicated by black circles (except for an inconsequential permutation 5-b).

communication, on the possibility that these quantum effects might lower the energy to such an extent that the logarithm in (24) should be absent. At any rate, although our energy estimates may be incorrect, quantized vortex lines probably exist. We continue our discussion of the consequences of this assumption.

On a large scale according to the theorem of Helmholtz, vorticity moves with the fluid in such a way that the strength of a vortex filament remains constant. This means that if the fluid drifts the lines drift with it, maintaining their quantized strength. This is true, at least, if no forces act directly on the vortex line. In general the force per unit length on a vortex line equals the density, $\varrho_0 m$, times the vector cross product of the circulation, $2\varrho\hbar m^{-1}$, and the velocity of fluid where the vortex is.

9. CRITICAL VELOCITY AND FLOW RESISTANCE

We next turn to the role such vortex lines may play in the resistance to flow found at sufficiently high velocities. We have suggested that this resistance cannot be understood in terms of a direct creation of rotons, the superfluid otherwise being in perfect flow. Let us consider what would happen if liquid is flowing out of an

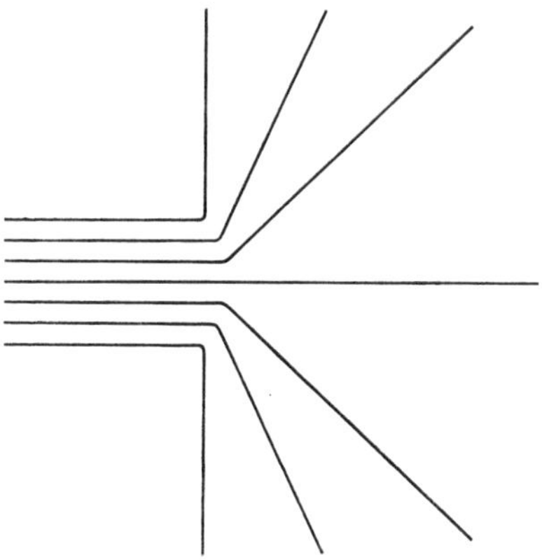

FIG. 7. Ideal potential flow from an orifice.

orifice, or tube, into a reservoir of fluid at rest. In Fig. 7 is illustrated the distribution of flow for irrotational motion. A very high velocity develops near the corners and large accelerations develop there. An ordinary fluid, such as water, flows in a complicated manner such as illustrated in Fig. 8 (a few moments after flow starts). The water shoots out straight into the nearly still fluid in the reservoir, forming a vortex sheet, which is unstable and curls around, eventually in an extremely complex manner. Let us see how helium might try to imitate some of the features of the type of flow illustrated in Fig. 8. Just for rough orientation and estimate suppose the fluid tries to go out in a jet, let us say at first of the

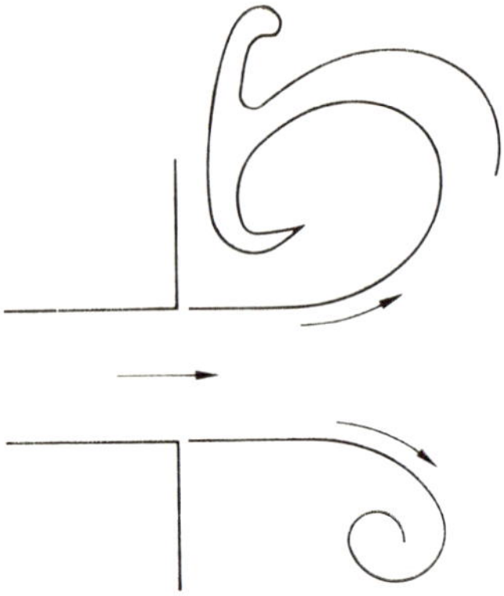

FIG. 8. Real flow from an orifice for ordinary liquids, producing an unstable vortex sheet.

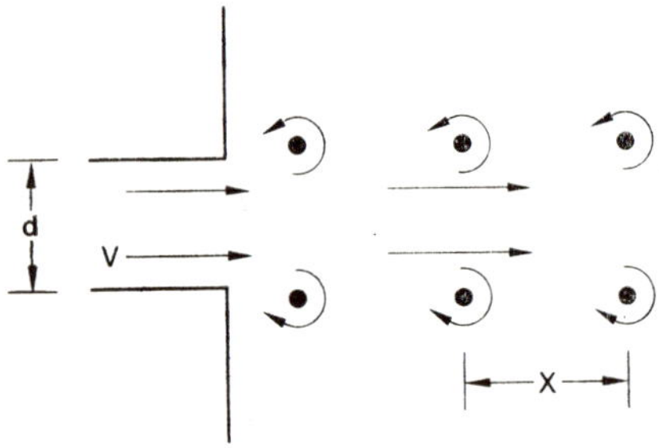

FIG. 9. Idealization of supposed vortex rings formed when superfluid helium issues at high speed from an orifice.

same width and velocity as in the tube is a long slot perpendicular to the paper, and the flow is roughly two dimensional.

Then circulation is implied for the velocity is v in the jet and 0 outside. This requires the formation of vortex lines, perhaps as illustrated in Fig. 9. The spacing is x and if this is small compared to d, the slot width, the velocity distribution is roughly uniform inside the jet. Taking a line integral along the jet for unit distance, and returning outside the jet, the circulation is v so the number of lines per centimeter is

$$\frac{1}{x} = v/2\pi\hbar m^{-1}.$$

It takes energy to form these lines. If there is not enough kinetic energy in the fluid to supply the energy to make the lines, no resistance will appear. Once the lines can be formed they are, in a manner we shall soon discuss, ultimately dissipated as heat and a resistance appears. Let us see what order of critical velocity we would estimate in this way. The lines move out at the velocity of the fluid at their own location, which is $v/2$. Another way to see the necessity for this is to realize that as the fluid passes from inside to outside the pipe vorticity is created, so new lines must continually come rolling out of the ends of the orifice. In our case v/x lines are created per second. The energy needed to create these is

$$\frac{v^2}{2\pi\hbar m^{-1}}\varrho_0\pi\hbar^2 m^{-1}\ln(d/a)$$

(per unit length of slot) where the argument in the logarithm is only approximate. The total kinetic energy available per cc of fluid is $\frac{m\varrho_o v^2}{2}$, so that per second $vd\frac{m\varrho_o v^2}{2}$ is available. If we define v_o as that velocity for which the energy available is just large enough to create the vortices we find

$$v_o d = \hbar m^{-1}\ln(d/a).$$

For example, for a slit of width $d = 10^{-5}$ cm, (which is about three times the width of a Rollin film at a height of 1 cm) this gives $v_o = 100$ cm/sec, if $\ln(d/a)$ is taken as 6. This is somewhat higher than the critical velocities observed. The calculation is only meant as an estimate because the actual situation must be complicated. For one thing, near the critical condition x comes out about $3d$ so our picture of a uniform jet is poor. Further, the velocity in the jet must of course be reduced as a result of the energy needed to form the vortex line. Actually probably the situation near the critical point must be very complicated and irregular. The flow for short momentary periods may be much like Fig. 7 but irregularly vortex lines peel off of the edges of the slit, probably starting at one point along the slit and progressing to other places, or perhaps if the hole is circular, one or two vortex lines are fed out continuously in a form roughly like a helix. It is predicted that very close to the critical velocity when loss just begins, the resistance will be irregular and show fluctuations. These fluctuations are very small, however, and would be hard to detect. Possibly some sound may be generated by the flow irregularities. It is difficult to estimate its intensity. When helium is driven, just above critical velocity, through an emery powder superleak, some noise should be generated as the various vortex lines suddenly form and pass into the stream. The irregularities are a result of the unpredictable quantum transitions between states of no vortex line and one with a section of line.

Another possible source of vortex lines is the contact between flowing liquid and the walls. It is not necessary that all the loss occurs at the exit end of the tube. The walls of the pipe are irregular. Vortex lines may be created inside the pipe also.

It is difficult to go beyond this order of magnitude calculation in describing the conditions controlling the production of vortex lines. For example, if one studies the example given there are serious difficulties. As a particular vortex line leaves the end of the tube there are very great forces trying to pull it back resulting from its image in the tube wall. Let us imagine a line a distance b above the wall in a tube in which the velocity of flow is v_o. It is readily shown that the forces acting on the line are these. First a force pulling away from the surface of strength $2\pi\hbar\varrho_o v_o$. Second, from the image, an attraction to the wall of strength $\pi\hbar^2\varrho_o/mb$. A vortex line responds to forces by moving through the liquid to reduce the net force to zero. In this case it would drift upstream if the attraction is highest. But a vortex line will interact with the wall, especially at its ends which go into the wall surface. Suppose this results in a frictional force which keeps the line from moving upstream. Then the response is to move closer to the wall. The vortex only moves away from the wall if $2\pi\hbar\varrho_o v$ exceeds $\pi\hbar^2\varrho_o/mb$. Even if v is 100 cm/sec this requires b to exceed 10^{-6} cm or 20 atomic spacings. We might expect a vortex line to fluctuate away from the surface by a few atomic diameters. But how can we expect to penetrate the enormous potential barrier, to create a line so far away from the surface that the flow velocity can pull it further out and create eventual vorticity and energy loss?

More likely a line gets started somehow and has its ends tied on the wall. Then the forces of the fluid on the rest of the line cause it to wander about in such a way that more and more vortex line is fed out. It is not necessary to create bodily at one instant a complete section of line. For example, for the case of liquid issuing from a tube perhaps the vortex lines are helices with contact point at the edge of the hole which turn round and round while the helix moves outward. Similar things could happen inside tubes. If the tubes are very narrow the line will hit the other surface easily and be attracted by the walls. It can never get very far from a wall. Even if started somehow it will fall back into the tube walls unless the velocity v_0 suffices to keep it in the stream. Therefore the smallest tubes have the highest critical velocities.

10. TURBULENCE

The patterns of vortex lines which we have studied are well known to be unstable. In the case of the rotating cylinder this is not true if the cylindrical vessel containing the helium rotates also. But if the container is stopped the situation is altered. There are forces between the wall and vortex lines. (This is because the fluid density is altered near the line axis, so the interaction with the wall is not the same as the average for the bulk of the helium). The lines at the outside drag past the stationary wall and as a result get distorted from their original vertical line position. This twists others, etc.

Lines fall into the wall and others twist about each other in a complex way. It would be interesting to study this experimentally, to see how fast, and in what manner, the liquid eventually slows down.

In ordinary fluids flowing rapidly and with very low viscosity the phenomenon of turbulence sets in. A motion involving vorticity is unstable. The vortex lines twist about in an ever more complex fashion, increasing their length at the expense of the kinetic energy of the main stream. That is, if a liquid is flowing at a uniform velocity and a vortex line is started somewhere upstream, this line is twisted into a long complex tangle further down stream. To the uniform velocity is added a complex irregular velocity field. The energy for this is supplied by pressure head.

We may imagine that similar things happen in the helium. Except for distances of a few ångströms from the core of the vortex, the laws obeyed are those of classical hydrodynamics. A single line playing out from points in the wall upstream (both ends of the line terminate on the wall, of course) can soon fill the tube with a tangle of line. The energy needed to form the extra length of line is supplied by a pressure head. (The force that the pressure head exerts on the lines acts eventually on the walls through the interaction of the lines with the walls). The resistance to flow somewhat above critical velocity must be the analogue in superfluid helium of turbulence, and a close analogue at that.

There are some ways, however, in which the two cases differ. In a classical fluid there is a thin boundary layer near the wall of the pipe in which viscosity controls the situation. In this boundary layer there is a large vorticity, but it escapes into the stream to be amplified, only from the edge of the layer. Inside it is damped by viscosity. As the stream velocity falls the boundary layer thickens, for the amplification is less and the damping overpowers it ever further from the wall. Below a critical velocity the turbulence ceases altogether and the flow is laminar, but with vorticity, the viscosity keeping the vorticity from amplifying itself. That is, viscosity is the mechanism which determines whether vorticity will be amplified or not, and therefore whether turbulence is produced. If the viscosity goes to zero as a limit (and no other physical phenomena are added) a classical ideal liquid would exhibit turbulence at any velocity, no matter how small.

Superfluid helium is an ideal fluid of zero viscosity. It does not exhibit turbulence at low velocity because of another, quantum mechanical, effect. The vorticity is quantized and cannot begin at as low amount as desired. One must supply energy enough to get the first one or two vortex lines started before the amplification process of turbulence can take over. There will not be a boundary layer with a structure analogous to that in classical flow (although near the walls the flow will be somewhat different because of the dragging forces between the moving vortex lines and the wall).

In a classical fluid, if the turbulent stream empties into a reservoir, the turbulent motion continues for a while, but as a result of the viscosity, it gradually slows up and dies out, the energy appearing eventually as heat.

What happens to a turbulent mass of superfluid left to itself? If there is normal fluid present the rotons and phonons will collide with the vortex lines and take energy from them, gradually turning this energy gain into more rotons and phonons (as a result of collisions among rotons the number of these may change). But an interesting question arises if the experiment is imagined at absolute zero. What can eventually become of the kinetic energy of the vortex lines?

One possibility that suggests itself is this. Consider a large distorted ring vortex (Fig. 10a). If, in a place, two oppositely directed sections of line approach closely, the situation is unstable, and the lines twist about each other in a complicated fashion, eventually coming very close; in places, nearly within an atomic spacing. Consider two such lines (Fig. 10b). With a small rearrangement, the lines (which are under tension) may snap together and join connections a new way to form two loops (Fig. 10c). Energy released this way goes into further twisting and winding of the new loops. This continues until the single loop has become chopped into a very large number of small loops (Fig. 10d). The smallest ring vortex that can exist must have a radius about half the atomic spacing. Let us guess that this is in fact a roton. Then all the energy

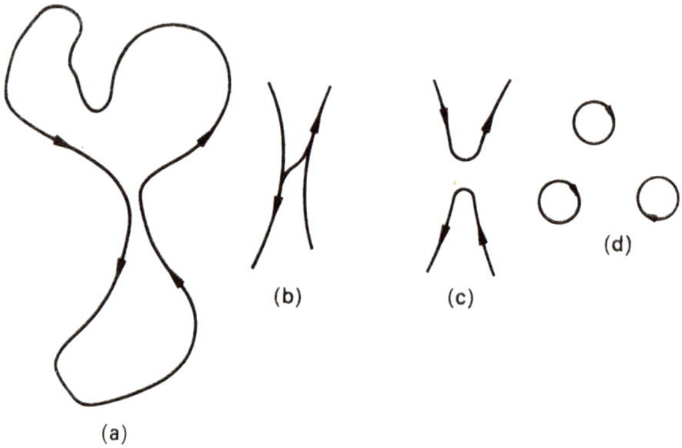

FIG. 10. A vortex ring (a) can break up into smaller rings if the transition between states (b) and (c) is allowed when the separation of vortex lines becomes of atomic dimensions. The eventual small rings (d) may be identical to rotons.

of the vortex will eventually end by forming large numbers of rotons, that is, heat. Perhaps eventually it will be easier to understand the details of the complete transformation of organized flow energy into disorganized heat energy for liquid helium than for other substances.

11. ROTONS AS RING VORTICES

It is not unreasonable to guess that these smallest vortices are rotons*. The velocity distribution around a roton, which is found by analytic means (ref. 8) is similar to that around a vortex ring.

* *(Note by Editor.)* For different treatments of rotons see de Boer (1963), Chester (1963) and Part 1 of the present volume.

It is quite reasonable that a vortex ring can be only so small. To increase the curvature of a vortex line beyond that of radius roughly a may take energy. Let us imagine a roton to be the circular quantized vortex of lowest energy. A large circular vortex has (from (21)) energy

$$E = 2\pi R \cdot \frac{\pi \hbar^2}{m} \varrho_0 \ln (R/a).$$

It carries momentum $p = \pi R^2 \cdot 2\pi\hbar\varrho_0$. This momentum is that of a roton, p_0, if $R = 2{\cdot}2$ Å. The energy is the right order (it corresponds to replacing ln by 1·6).

One might object that such a vortex drifts through the fluid, at velocity $v = (\hbar/2mR) \ln (R/a)$, so one would expect rotons not to have a zero group velocity. Actually this drift, of a large vortex, has its seat in the force tending to shrink the vortex to decrease the energy of the line. The response to the radially directed force is a perpendicular motion. It is analogous to the ornery response of a gyroscope. In fact, if a vortex line were a thin flexible mechanical tube with inertia, and were started with zero forward motion, it would first fall in a bit and then move forward in a halting fashion, like the nutation of a gyroscopic, or the motion of an electron in crossed magnetic and electric fields. In a roton we imagine that the forces tending to contract the ring are already opposed by a kind of stiffness of the ring. It is already as small as possible. No drift motion results. In fact forward drift would expand it and raise the energy, while reverse drift would try to compress it to smaller size, again raising the energy. The lowest energy is at zero drift velocity. We may notice in passing that they can only drift in a direction perpendicular to their plane, that is, along, or opposite, the direction of the momentum. This agrees with a property derived for rotons from their energy-momentum relation (1), that the group velocity $\delta E(p)/\delta p$ is in the direction of the momentum (or opposite).

Having travelled so far making one unverified conjecture upon another we may have strayed very far from the truth. However imprudent it may be, there is one further observation we would like to make. A detailed picture is not available which describes physically just what goes on as the transition is approached from below. The free energy expression arising from (6) does not of itself describe the transition. The transition occurs when the number of rotons is very large. Some sort of interaction may occur between them, or there may be some limitation to the degrees of freedom.(18) There is no doubt that it is the analogue of the transition in the ideal gas, but it would be nice if we could get a less mathematical and formal description of the events. Of the following I am not sure, but it does seem to be an interesting possibility.

If rotons are the smallest ring vortices, and those of lowest energy, Δ, then there are states of higher energy corresponding to larger rings. For example, a ring of twice the diameter may have twice the energy more or less. The relative number of these will be expected to be very low, however. Since Δ is 9·6°K, at the transition $\exp(-\Delta/kT)$ is 10^{-2}, so very few larger vortices will be expected in equilibrium. Certainly none whose length is 10^2 or 10^3 atoms! This neglects an important feature, however. For a long line there are an enormous number of shapes and orientations available. Such a line is not infinitely flexible, of course, for the curvature cannot well exceed a^{-1}. It may be likened to a chain of a finite number of links. Adding one link requires an energy ε, say of order Δ, but increases the number of orientations by some factor, asymptotically, say s. In equilibrium then, the number of chains of $n+1$ links is a factor $s \exp(-\varepsilon/kT)$ times the number with n links. For low temperatures this is less than unity. No long chains are important. The excitations consist of rotons and a few other rings of slightly larger size. As the temperature rises, however, there comes a time when the factor $s \exp(-\varepsilon/kT)$ exceeds unity. Then suddenly the rings of very largest length are of importance. The state with one vortex line (or a very few) which winds and winds throughout the liquid like a near approximation to a Jordon curve, is no longer of negligible weight. The superfluid is pierced through and through with vortex line. We are describing the disorder of Helium I. At first the curve doesn't make full use of all of its orientations and higher entropy. But as the temperature rises a little more it squeezes into the last corners and pockets of superfluid until it has no more degrees of flexibility available. The specific heat curve drops off from the transition to a smooth curve and the memory of the possibility the helium can exhibit quantum properties in a unique way is lost in the perfusion of states and in disorder, as it is for more usual liquids.

REFERENCES

1. L. Onsager, *Nuov. Cim.*, **6,** Supp., **2,** 249 (1949).
2. F. London, *Nature*, **141,** 643 (1938).
3. A. Einstein, *Ber. Berl. Akad.*, **3** (1925).
4. F. London, *Phys. Rev.*, **54,** 947 (1938).
5. R. P. Feynman, *Phys. Rev.*, **91,** 1291 (1953).
6. G. V. Chester, *Phys. Rev.*, **94,** 246 (1954).
7. R. P. Feynman, *Phys. Rev.*, **91,** 1301 (1953).
8. R. P. Feynman, *Phys. Rev.*, **94,** 262 (1954).
9. L. Tisza, *Phys. Rev.*, **72,** 838 (1947).
10. L. Landau, *J. Phys. U.S.S.R.*, **5,** 71 (1941).
11. R. B. Dingle, *Advances in Phys.*, **1,** 112 (1952).
12. D. de Klerk, R. P. Hudson and J. R. Pellam, *Phys. Rev.*, **93,** 28 (1954).
13. L. Landau and I. Khalatnikov, *J. Exper. Theor. Phys. (U.S.S.R.)*, **19, 637,** 709 (1949).
14. J. C. Slater and J. G. Kirkwood, *Phys. Rev.*, **37,** 682 (1932).
15. A. Bijl, *Leiden, Comm. Suppl.*, **90a,** *Physica*, **7,** 869 (1946).
16. D. G. Henshaw and D. C. Hurst, *Phys. Rev.*, **91,** 1222 (1953).
17. J. Reekie and T. S. Hutchinson, *Phys. Rev.*, **92,** 827 (1953). We are grateful to Dr. Reekie for sending us the curve of $S(k)$ derived from his data.
18. R. P. Feynman, *Phys. Rev.* **94,** 262 (1954), especially the discussion preceding expression (28). There is a typographical error there. The relation (28) should read as an inequality, the left side not greater than the right side.

T6. KINETIC COEFFICIENTS OF HELIUM II*

I. M. KHALATNIKOV

Summary. Starting from the kinetic equation for the excitations in helium II we derive the equations of motion of helium II. We elucidate the problem of the kinetic coefficients in helium II. We obtain general expressions which determine these coefficients: the coefficient of first viscosity, the three coefficients of second viscosity and, finally, a coefficient which is the equivalent of the heat-conductivity coefficient in ordinary solids.

A detailed analysis of the kinetic equation for the elementary excitations in helium II has enabled us to elucidate the problem of its kinetic coefficients, it was then seen that apart from the coefficient of first viscosity and the three coefficients of second viscosity[1] there is in helium II yet another kinetic coefficient which is analogous to the heat conductivity coefficient in ordinary solids. This coefficient connects the temperature gradient with the heat current. We start our considerations with an analysis of the kinetic equation for the elementary excitations in helium II. This analysis allows us to obtain the hydrodynamic equations for helium II and the general expressions for the energy and momentum currents.

1. KINETIC EQUATION FOR THE ELEMENTARY EXCITATIONS IN HELIUM II

We can write the kinetic equation which determines the distribution function n of the elementary excitations in helium II in the form of a continuity equation in the space of the coordinates

* *J. Exptl. Theoret. Phys. (U.S.S.R.)* **23**, 8 (1952).

$\boldsymbol{r}$ and momenta $\boldsymbol{p}$:

$$\dot{n}+\left(\frac{\partial}{\partial \boldsymbol{r}}\cdot n\dot{\boldsymbol{r}}\right)+\left(\frac{\partial}{\partial \boldsymbol{p}}\cdot n\dot{\boldsymbol{p}}\right) = J(n), \tag{1.1}$$

where $J(n)$ is the collision integral, the exact form of which is of no interest to us in the present section.

The velocity $\dot{\boldsymbol{r}}$ and the force $\dot{\boldsymbol{p}}$ can according to the Hamiltonian equations of motion be expressed in terms of the Hamiltonian H of the elementary excitations:

$$\dot{\boldsymbol{r}} = \frac{\partial H}{\partial \boldsymbol{p}}, \quad \dot{\boldsymbol{p}} = -\frac{\partial H}{\partial \boldsymbol{r}}. \tag{1.2}$$

Substituting (1.2) into (1.1) and expanding the derivatives of the products $n\dot{\boldsymbol{r}}$ and $n\dot{\boldsymbol{p}}$, we can write (1.1) in the form

$$\dot{n}+\left(\frac{\partial n}{\partial \boldsymbol{r}}\cdot\frac{\partial H}{\partial \boldsymbol{p}}\right)-\left(\frac{\partial n}{\partial \boldsymbol{p}}\cdot\frac{\partial H}{\partial \boldsymbol{r}}\right) = J(n). \tag{1.3}$$

Let there be a superfluid motion in helium II with velocity $\boldsymbol{v}_s$. The Hamiltonian is then equal to

$$H = \varepsilon(p)+(\boldsymbol{p}\cdot\boldsymbol{v}_s), \tag{1.4}$$

where $\varepsilon(p)$ is the excitation energy in a system of reference which is moving with velocity $\boldsymbol{v}_s$.

Equations of motion of helium II

Let us multiply the left-hand and right-hand sides of equation (1.3) by the ith component of the momentum, p_i, and integrate them over the whole of momentum space. Because of the conservation of the momentum of the whole system, the integral of the right-hand side vanishes:

$$\int p_i J(n)\, d\tau_{\boldsymbol{p}} = 0 \tag{1.5}$$

We thus have

$$\frac{\partial}{\partial t}\int np_i d\tau_{\boldsymbol{p}}+\int p_i\frac{\partial H}{\partial p_k}\frac{\partial n}{\partial r_k}d\tau_{\boldsymbol{p}}-\int p_i\frac{\partial H}{\partial r_k}\frac{\partial n}{\partial p_k}d\tau_{\boldsymbol{p}} = 0.$$

Furthermore, we get by some simple transformations

$$\frac{\partial}{\partial t}\int np_i\,d\tau_p+\frac{\partial}{\partial r_k}\int np_i\frac{\partial H}{\partial p_k}\,d\tau_p+\int n\frac{\partial H}{\partial r_i}\,d\tau_p=0. \quad (1.6)$$

If we then substitute the expression (1.4) for H and for the sake of simplicity indicate the integration over $\boldsymbol{p}$-space by a bar over the corresponding quantity we have

$$\frac{\partial}{\partial t}\overline{np_i}+\frac{\partial}{\partial r_k}\overline{np_i\left(\frac{\partial \varepsilon}{\partial p_k}+v_{sk}\right)}+\overline{n\left(\frac{\partial \varepsilon}{\partial r_i}+p_k\frac{\partial v_{sk}}{\partial r_i}\right)}=0. \quad (1.7)$$

Equation (1.7) is one of the equations of motion of liquid helium II. It determines the acceleration of the relative motion of the normal and the superfluid parts of helium II. This follows directly from the fact that $\overline{\boldsymbol{np}}$ is the total momentum of the excitations in the system of reference which corresponds to the superfluid motion (the superfluid part of the liquid is at rest).

We can express the total momentum per unit volume of helium II in terms of the momentum of the excitations $\overline{\boldsymbol{np}}$ in the system of reference moving with a velocity $\boldsymbol{v}_s$ using the relation $\boldsymbol{j}=\overline{\boldsymbol{np}}+\varrho\boldsymbol{v}_s$. The derivative with respect to time of the total momentum is, as a consequence of the general conservation laws for energy and momentum, equal to the divergence of a symmetric tensor Π_{ik}. The momentum flux tensor Π_{ik} in the fixed system of reference can be expressed in terms of its value π_{ik} in the system of reference which is moving with the velocity $\boldsymbol{v}_s$.*

$$\Pi_{ik}=\pi_{ik}+v_{sk}\overline{np_i}+v_{si}\overline{np_k}+\varrho v_{si}v_{sk}. \quad (1.8)$$

* One can easily obtain the connection between the value of the momentum flux tensor in the fixed system of reference(Π_{ik}) and that in the system of reference moving with a velocity $\boldsymbol{v}$(π_{ik}) in the case of classical hydrodynamics where the form of the tensor Π_{ik} is known. One can write the tensor Π_{ik} of classical hydrodynamics in the form

$$\Pi_{ik}=\varrho u_i u_k+p\delta_{ik}$$

[*footnote continued on next page*

Thus,

$$\frac{\partial}{\partial t}\overline{np_i}+\frac{\partial}{\partial t}\varrho v_{si}+\frac{\partial}{\partial x_k}(\pi_{ik}+v_{sk}\overline{np_i}+v_{si}\overline{np_k}+\varrho v_{si}v_{sk})=0, \quad (1.9)$$

we subtract equation (1.7) from (1.9) and use the equation of continuity

$$\frac{\partial}{\partial t}\varrho+\frac{\partial}{\partial r_k}(\overline{np_k}+\varrho v_{sk})=0. \quad (1.10)$$

As a result we get

$$\varrho\frac{\partial}{\partial t}v_{si}+\varrho v_{sk}\frac{\partial v_{si}}{\partial r_k}+\frac{\partial}{\partial r_k}\pi_{ik}-n\overline{\frac{\partial\varepsilon}{\partial r_i}}-\frac{\partial}{\partial r_k}\overline{np_i\frac{\partial\varepsilon}{\partial p_k}}=0. \quad (1.11)$$

When there is no external field the derivative $\partial\varepsilon/\partial r_i$ is equal to $(\partial\varepsilon/\partial\varrho)(\partial\varrho/\partial r_i)$. From the condition curl $\boldsymbol{v}_s=0$ follows the requirement that the sum of the last three terms in (1.11) should be the product of the density ϱ and the gradient of some function. Moreover, it is absolutely obvious that the tensor π_{ik} at the absolute zero when there are no excitations must be equal to $p_0\delta_{ik}$ (p_0 is the pressure of liquid helium II without excitations).

These requirements uniquely determine the form of the tensor π_{ik}:

$$\pi_{ik}=\overline{np_i\frac{\partial\varepsilon}{\partial p_k}}+\delta_{ik}\overline{n\frac{\partial\varepsilon}{\partial p}}\varrho+\delta_{ik}p_0. \quad (1.12)$$

($\boldsymbol{u}$ is the velocity of the liquid, p the pressure). We now express the velocity $\boldsymbol{u}$ of the liquid in the fixed system of reference in terms of the velocity $\boldsymbol{u}'$ in the system of reference moving with the velocity $\boldsymbol{v}$:

$$\boldsymbol{u}=\boldsymbol{u}'+\boldsymbol{v}.$$

After substituting this expression for $\boldsymbol{u}$ into the above formula for Π_{ik} we get

$$\Pi_{ik}=\varrho v_i v_k+\varrho u'_i v_k+\varrho u'_k v_i+(\varrho u'_i u'_k+p\delta\ ).$$

Denoting the momentum per unit volume of liquid in the moving system of reference by $\boldsymbol{p}$ we finally obtain the required relation:

$$\Pi_{ik}=\varrho v_i v_k+p_i v_k+p_k v_i+\pi_{ik}$$

It is natural to assume that the formula which we have obtained for the transformation of the tensor Π_{ik} retains its form also in the hydrodynamics of helium II.

The pressure p_0 can be expressed in terms of the energy E_0 and the potential Φ_0 ($T = 0$):

$$p_0 = -E_0+\Phi_0\varrho \quad \left(\Phi_0 = \frac{\partial E_0}{\partial \varrho},\ T = 0\right).$$

Hence it follows that

$$\frac{\partial p_0}{\partial r_i} = \varrho \frac{\partial \Phi_0}{\partial r_i}\,. \tag{1.13}$$

We substitute expression (1.12) for π_{ik} into (1.11); using (1.13) we get

$$\varrho \frac{\partial}{\partial t} v_{si}+\varrho v_{sk}\frac{\partial v_{si}}{\partial r_k}+\varrho\frac{\partial}{\partial r_k}\overline{n\frac{\partial \varepsilon}{\partial p}}+\varrho\frac{\partial}{\partial r_i}\Phi_0 = 0. \tag{1.14}$$

Dividing by the common factor ϱ in (1.14) we find finally

$$\frac{\partial}{\partial t} v_s+\nabla\left\{\Phi_0+n\frac{\partial \varepsilon}{\partial \varrho}+\frac{v_s^2}{2}\right\} = 0. \tag{1.15}$$

The equation obtained is the equation for the superfluid motion. The total thermodynamic potential Φ is according to (1.15) equal to

$$\Phi = \Phi_0+\int n\frac{\partial \varepsilon}{\partial \varrho}\, d\tau_p. \tag{1.16}$$

The second term in (1.16) is due to the contribution of the excitations to the thermodynamic potential Φ.

It is sometimes convenient to use the equation of motion (1.9) instead of equation (1.7); this equation takes after substituting into it expression (1.12) for the tensor π_{ik} the form

$$\frac{\partial}{\partial t}(\overline{np_i}+\varrho v_{si})$$

$$+\frac{\partial}{\partial x_k}\left\{\overline{np_i\frac{\partial \varepsilon}{\partial p_k}}+\delta_{ik}\overline{n\frac{\partial \varepsilon}{\partial \varrho}}\varrho+\delta_{ik}p_0+\overline{np_i}v_{ik}+\overline{np_k}v_{si}+\varrho v_{si}v_{sk}\right\} = 0. \tag{1.9'}$$

Energy conservation law

The energy E per unit volume of liquid helium II in the fixed system of reference is according to the well-known transformation formulae connected with the energy E_{int} in the system of reference moving with a velocity $\boldsymbol{v}_s$ by the relation

$$E = \frac{\varrho v_s^2}{2} + (v_s \cdot \overline{np}) + E_{\text{int}}. \tag{1.17}$$

If we differentiate the first two terms in (1.17), which give the kinetic energy of helium (subtracting the energy of the relative motion of the normal and the superfluid parts), with respect to the time and use the equations of motion (1.7) and (1.15) and the equation of continuity (1.10) we get

$$\frac{\partial E_{\text{kin}}}{\partial t} = \frac{\partial}{\partial t}\left(\frac{\varrho v_s^2}{2} + (v_s \cdot \overline{np})\right) = -(\varrho v_{si} + \overline{np_i})\frac{\partial}{\partial r_i}\left(\phi_0 + \overline{n\frac{\partial \varepsilon}{\partial \varrho}} + \frac{v_s^2}{2}\right)$$
$$-\frac{v_s^2}{2}\frac{\partial}{\partial r_i}(\varrho v_{si} + \overline{np_i}) - v_{si}\frac{\partial}{\partial r_k}\overline{np_i\left(\frac{\partial \varepsilon}{\partial p_k} + v_{sk}\right)} - v_{si}\,\overline{n\left(\frac{\partial \varepsilon}{\partial p_i}n + p_k\frac{\partial v_{sk}}{\partial r_i}\right)}. \tag{1.18}$$

We find now the time derivative of the internal energy of helium $E_{\text{int}} = \overline{n\varepsilon}$. To do this we multiply the left-hand and right-hand sides of the kinetic equation (1.3) by ε and integrate it over the whole of $\boldsymbol{p}$-space. The integral of the right-hand side $\int J(n)\varepsilon\, d\tau_p$ vanishes due to the energy conservation law. As a result we get

$$\overline{\varepsilon\frac{\partial n}{\partial t}} + \overline{\varepsilon\frac{\partial n}{\partial r_s}\frac{\partial H}{\partial p_k}} - \overline{\varepsilon\frac{\partial n}{\partial p_k}\frac{\partial H}{\partial r_k}} = 0. \tag{1.19}$$

Then, using some simple tranformations of the equation obtained and using Gauss's theorem, we find

$$\frac{\partial}{\partial t}\overline{\varepsilon n} + \frac{\partial}{\partial r_k}\overline{\varepsilon n\frac{\partial H}{\partial p_k}} - \overline{n\frac{\partial}{\partial r_k}\left(\varepsilon\frac{\partial H}{\partial p_k}\right)} - \overline{\frac{\partial}{\partial p_k}\varepsilon n\frac{\partial H}{\partial r_k}} + \overline{n\frac{\partial}{\partial p_k}\varepsilon\frac{\partial H}{\partial r_k}} - \overline{n\frac{\partial \varepsilon}{\partial \varrho}}\frac{\partial \varrho}{\partial t}$$
$$= \frac{\partial}{\partial t}\overline{\varepsilon n} + \frac{\partial}{\partial r_k}\overline{\varepsilon n\left(\frac{\partial \varepsilon}{\partial p_k} + v_{sk}\right)} - v_{sk}\overline{n\frac{\partial \varepsilon}{\partial r_k}} + \overline{n\frac{\partial \varepsilon}{\partial p_k}p_i}\frac{\partial v_{si}}{\partial r_k} - \overline{n\frac{\partial \varepsilon}{\partial \varrho}}\frac{\partial \varrho}{\partial t} = 0. \tag{1.20}$$

We thus have

$$\frac{\partial}{\partial t}E_{\text{int}} = \frac{\partial}{\partial t}\overline{(n\varepsilon)} = -\frac{\partial}{\partial r_k}\overline{\varepsilon n\left(\frac{\partial\varepsilon}{\partial p_k}+v_{sk}\right)}$$

$$+\overline{n\frac{\partial\varepsilon}{\partial r_k}}v_{sk}-\overline{n\frac{\partial\varepsilon}{\partial p_k}p_i}\frac{\partial v_{si}}{\partial r_k}-\overline{n\frac{\partial\varepsilon}{\partial\varrho}}\frac{\partial\varrho}{\partial t}. \tag{1.21}$$

The derivative of the sum of the kinetic and the internal energy of helium II is according to (113) and (1.21) equal to

$$\frac{\partial}{\partial t}(E_{\text{kin}}+E_{\text{int}}) = -\frac{\partial}{\partial r_k}\left\{(\overline{np_k}+\varrho v_{sk})\left(\frac{v_s^2}{2}+\overline{n\frac{\partial\varepsilon}{\partial\varrho}}+\phi_0\right)\right\}$$

$$+\left(\overline{n\frac{\partial\varepsilon}{\partial\varrho}}+\phi_0\right)\frac{\partial}{\partial r_k}(\overline{np_k}+\varrho v)_{sk})-\frac{\partial}{\partial r_k}\overline{\varepsilon n\left(\frac{\partial\varepsilon}{\partial p_k}+v_{sk}\right)}$$

$$-v_{si}\frac{\partial}{\partial r_k}\overline{np_i\left(\frac{\partial\varepsilon}{\partial p_k}+v_{sk}\right)}-\frac{\partial v_{si}}{\partial r_k}\overline{np_i\left(\frac{\partial\varepsilon}{\partial p_k}+v_{sk}\right)}-\overline{n\frac{\partial\varepsilon}{\partial\varrho}}\frac{\partial\varrho}{\partial t}.$$

Hence we get after simple transformations and using equation (1.10) the energy conservation law in the form of an equation of continuity for the total energy:

$$\frac{\partial}{\partial t}E = E_{\text{kin}}+E_{\text{int}}+\phi_0\frac{\partial\varrho}{\partial t}$$

$$= -\frac{\partial}{\partial r_k}\left\{(\overline{np_k}+\varrho v_{sk})\left(\phi_0+\overline{n\frac{\partial\varepsilon}{\partial\varrho}}+\frac{v_s^2}{2}\right)+\overline{nH\frac{\partial}{\partial p_k}H}\right\}. \tag{1.22}$$

The third term on the left-hand side of (1.22) characterizes the change of energy of liquid helium II without excitations. The energy flux $\boldsymbol{Q}$ in helium II is according to (1.22) equal to

$$\boldsymbol{Q} = (\overline{n\boldsymbol{p}}+\varrho\boldsymbol{v}_s)\left(\phi_0+\overline{n\frac{\partial\varepsilon}{\partial\varrho}}+\frac{v_s^2}{2}\right)+\overline{nH\frac{\partial}{\partial p_k}H}. \tag{1.23}$$

Equations (1.7), (1.10), (1.15), together with equation (1.22) are a complete system of hydrodynamic equations for helium II. One usually writes down instead of the equation of continuity for the energy (1.22) the equation of continuity for the entropy which follows from (1.22) and the thermodynamic identities.

2. KINETIC COEFFICIENTS*

The equilibrium distribution function for excitations in helium II in which we have uniform normal motion with a velocity $\boldsymbol{v}_n$ and a uniform superfluid motion with a velocity $\boldsymbol{v}_s$ has the form (Bose statistics)

$$n_0 = (e^{(H-(p\cdot v_n))/kT}-1)^{-1} = (e^{(\varepsilon-(p\cdot v_n)+(p\cdot v_s))/kT}-1)^{-1}. \quad (2.1)$$

The equilibrium distribution function n_0 satisfies the kinetic equation making its right-hand side vanish. In the case where the system of excitations in helium II differs from equilibrium, the distribution of the excitations is determined by another function, the form of which can, however, in the general case not be given, as the kinetic equation can not be solved in the general case. The problem simplifies for the case when the spatial derivatives of the quantities determining the state of the system are small, that is, these quantities themselves change little over distances which characterize the size of the system. In that case the distribution function n differs little from the equilibrium function n_0. This fact makes it possible to linearize the kinetic equation. To do this we put in the left-hand side of the kinetic equation the equilibrium function n_0; we retain in it after differentiation only the terms which are linear in the spatial derivatives. The right-hand side of the kinetic equation (collision integral) is expanded in a series in the difference $n-n_0$. We thus obtain a linear integral equation which determines the difference $n-n_0$. We thus obtain a linear integral equation which determines the difference $n-n_0$ in a form proportional to the spatial derivatives of the quantities characterizing the state. The values of $n-n_0$ found in this way allow us to evaluate the energy dissipation in a non-equilibrium system and to determine the form of the kinetic coefficients.

* *(Note by Editor.)* Hohenberg and Martin (1965) have given expressions for the kinetic coefficients in terms of correlation functions (Kubo-type formulae).

We elucidate in general form the properties of the kinetic coefficients in helium II. We substitute the distribution function n_0 into the left-hand side of the kinetic equation. We shall here assume that not only the spatial derivatives but also the velocity difference $\boldsymbol{v}_n - \boldsymbol{v}_s$ is small. This fact in no way restricts the problem as the velocity difference $\boldsymbol{v}_n - \boldsymbol{v}_s$ must be small compared to the sound velocity in helium II. It is well known that superfluidity breaks down appreciably earlier than this limit can be reached. We shall write down separately each of the terms contained on the left-hand side of the kinetic equation:*

$$\frac{\partial n_0}{\partial t} = n' \left\{ \frac{1}{kT} \left(\frac{\partial \varepsilon}{\partial t} - \left(\boldsymbol{p} \cdot \frac{\partial \boldsymbol{v}_n}{\partial t} \right) + \left(\boldsymbol{p} \cdot \frac{\partial \boldsymbol{v}_s}{\partial t} \right) \right) - \frac{1}{kT^2} (\varepsilon - (\boldsymbol{p} \cdot \boldsymbol{v}_n) + (\boldsymbol{p} \cdot \boldsymbol{v}_s)) \frac{\partial T}{\partial t} \right\}. \tag{2.3}$$

We choose the density and the entropy as our independent variables. According to the hydrodynamical equations the time derivatives can be expressed in terms of the spatial derivatives in the linear approximation as follows:

$$\begin{aligned} \partial \varrho / \partial t &= -\operatorname{div} \boldsymbol{j}; \\ \partial S / \partial t &= -S \operatorname{div} \boldsymbol{v}_n; \\ \varrho_n (\partial / \partial t)(\boldsymbol{v}_n - \boldsymbol{v}_s) &= -S \nabla T \end{aligned} \tag{2.4}$$

($\boldsymbol{j} = \varrho_n \boldsymbol{v}_n + \varrho_s \boldsymbol{v}_s$ is the helium II current, and S the entropy per cm^3 of helium II).

Using (2.4) we can transform (2.3) to

$$\begin{aligned} \frac{\partial n_0}{\partial t} = \frac{n'}{kT} \Bigg\{ & \operatorname{div} (\boldsymbol{j} - \varrho \boldsymbol{v}_n) \left[\frac{1}{T} \frac{\partial T}{\partial \varrho} (\varepsilon - (\boldsymbol{p} \cdot \boldsymbol{v}_n) + (\boldsymbol{p} \cdot \boldsymbol{v}_s)) - \frac{\partial \varepsilon}{\partial \varrho} \right] \\ & + \operatorname{div} \boldsymbol{v}_n \left[\frac{1}{T} \left(\frac{\partial T}{\partial \varrho} \varrho + \frac{\partial T}{\partial S} S \right) (\varepsilon - (\boldsymbol{p} \cdot \boldsymbol{v}_n) + (\boldsymbol{p} \cdot \boldsymbol{v}_s) - \frac{\partial \varepsilon}{\partial \varrho} \varrho \right] \\ & + \frac{ST}{\varrho_n} \left(\boldsymbol{p} \cdot \frac{\nabla T}{T} \right) - (\nabla \varepsilon \boldsymbol{v}_n) + \left(\frac{\nabla T}{T} \boldsymbol{v}_n \right) (\varepsilon - (\boldsymbol{p} \cdot \boldsymbol{v}_n) + (\boldsymbol{p} \cdot \boldsymbol{v}_s)) \Bigg\}. \end{aligned} \tag{2.5}$$

* We indicate by a prime the differentiation of a function with respect to the argument

$$(\varepsilon - (\boldsymbol{p} \cdot \boldsymbol{v}_n) + (\boldsymbol{p} \cdot \boldsymbol{v}_s))/kT. \tag{2.2}$$

Moreover, we have

$$\left(\frac{\partial n}{\partial \boldsymbol{r}}\cdot\frac{\partial H}{\partial \boldsymbol{p}}\right)-\left(\frac{\partial n}{\partial \boldsymbol{p}}\cdot\frac{\partial H}{\partial \boldsymbol{r}}\right) = \frac{n'}{kT}\left\{-\left(\left(\frac{\partial \varepsilon}{\partial \boldsymbol{p}}+\boldsymbol{v}_s\right)\cdot\nabla(\boldsymbol{p}\cdot\boldsymbol{v}_n)\right)\right.$$
$$\left.-\left(\frac{\nabla T}{T}(\varepsilon-(\boldsymbol{p}\cdot\boldsymbol{v}_n)+(\boldsymbol{p}\cdot\boldsymbol{v}_s))\right)\cdot\left(\frac{\partial \varepsilon}{\partial \boldsymbol{p}}+\boldsymbol{v}_s\right)\right)+(\boldsymbol{v}_n\cdot\nabla[(\boldsymbol{p}\cdot\boldsymbol{v}_s)+\varepsilon])\right\}. \tag{2.6}$$

We substitute (2.5) and (2.6) into the left-hand side of the kinetic equation (1.1); collecting similar terms we get

$$\frac{n'}{kT}\left\{\left[\frac{1}{T}\frac{\partial T}{\partial \varrho}(\varepsilon-(\boldsymbol{p}\cdot\boldsymbol{v}_n)+(\boldsymbol{p}\cdot\boldsymbol{v}_s))-\frac{\partial \varepsilon}{\partial \varrho}\right]\operatorname{div}(\boldsymbol{j}-\varrho\boldsymbol{v}_n)\right.$$
$$+\left[\frac{1}{T}\left(\frac{\partial T}{\partial \varrho}\varrho+\frac{\partial T}{\partial S}S\right)(\varepsilon-(\boldsymbol{p}\cdot\boldsymbol{v}_n)+(\boldsymbol{p}\cdot\boldsymbol{v}_s))-\frac{\partial \varepsilon}{\partial \varrho}\varrho\right]\operatorname{div}\boldsymbol{v}_n$$
$$+\left(\frac{\nabla T}{T}\cdot\left[\boldsymbol{p}\frac{ST}{\varrho_n}-(\varepsilon-(\boldsymbol{p}\cdot\boldsymbol{v}_n)+(\boldsymbol{p}\cdot\boldsymbol{v}_s))\left(\frac{\partial \varepsilon}{\partial \boldsymbol{p}}-\boldsymbol{v}_n+\boldsymbol{v}_s\right)\right]\right)$$
$$\left.-\left(\frac{\partial \varepsilon}{\partial \boldsymbol{p}}\cdot\nabla(\boldsymbol{p}\cdot\boldsymbol{v}_n)-(\boldsymbol{v}_s\cdot\nabla(\boldsymbol{p}\cdot\boldsymbol{v}_n))+(\boldsymbol{v}_n\nabla(\boldsymbol{p}\cdot\boldsymbol{v}_s)\right)\right\} = J(n). \tag{2.7}$$

The last two terms in the braces in (2.7) can be neglected as their difference can easily be shown to be of the order of $(\boldsymbol{v}_n-\boldsymbol{v}_s)^2$. All terms containing the factor $(\boldsymbol{p}\cdot\boldsymbol{v}_n)-(\boldsymbol{p}\cdot\boldsymbol{v}_s)$ also turn out to be of the same order as $(\boldsymbol{v}_n-\boldsymbol{v}_s)^2$. We neglect them also. Finally we have

$$\frac{n'}{kT}\left\{\left(\frac{\varepsilon}{T}\frac{\partial T}{\partial \varrho}-\frac{\partial \varepsilon}{\partial \varrho}\right)\operatorname{div}(\boldsymbol{j}-\varrho\boldsymbol{v}_n)+\left[\frac{\varepsilon}{T}\left(\frac{\partial T}{\partial \varrho}\varrho+\frac{\partial T}{\partial S}S\right)\right.\right.$$
$$\left.-\frac{\partial \varepsilon}{\partial \varrho}\varrho\right]\operatorname{div}v_n+\left(\frac{\nabla T}{T}\cdot\left[\boldsymbol{p}\frac{ST}{\varrho_n}-\varepsilon\frac{\partial \varepsilon}{\partial \boldsymbol{p}}\right]\right)$$
$$\left.-\left(\frac{\partial \varepsilon}{\partial \boldsymbol{p}}\cdot\nabla(\boldsymbol{p}\cdot\boldsymbol{v}_n)\right)\right\} = J(n), \tag{2.8}$$

$$n' = -n_0(n_0+1), \quad n_0 = (e^{\varepsilon/kT}-1)^{-1},$$

We can write the solution of (2.8) in the form

$$n-n_0 = \frac{n'}{kT}\left\{A\left[\frac{\varepsilon}{T}\frac{\partial T}{\partial \varrho}-\frac{\partial \varepsilon}{\partial \varrho}\right]\operatorname{div}(\boldsymbol{j}-\varrho \boldsymbol{v}_n)\right.$$

$$+A\left[\frac{1}{T}\left(\frac{\partial T}{\partial \varrho}\varrho+\frac{\partial T}{\partial S}S\right)\varepsilon-\frac{\partial \varepsilon}{\partial \varrho}\varrho-\frac{1}{3}\left(\frac{\partial \varepsilon}{\partial \boldsymbol{p}}\cdot\boldsymbol{p}\right)\right]\operatorname{div}\boldsymbol{v}_n$$

$$+B\left(\frac{\nabla T}{T}\cdot\left[\boldsymbol{p}\frac{ST}{\varrho_n}-\varepsilon\frac{\partial \varepsilon}{\partial \boldsymbol{p}}\right]\right)+C\left[\frac{\partial \varepsilon}{\partial p_i}-p_k\frac{1}{3}\delta_{ik}\left(\frac{\partial \varepsilon}{\partial \boldsymbol{p}}\cdot\boldsymbol{p}\right)\right]\left(\frac{\partial v_{nk}}{\partial r_i}\right.$$

$$\left.\left.+\frac{\partial v_{ni}}{\partial r_k}-\frac{2}{3}\delta_{ik}\frac{\partial v_{nl}}{\partial r_l}\right)\right\}, \tag{2.9}$$

where A, B, and C are some state functions the form of which is determined by the nature of the interactions between the excitations.* The coefficient A is common to all scalar type terms, the coefficient B to the vector terms (terms proportional to $\boldsymbol{p}$) and, finally, the coefficient C to the tensor terms. From the last term within the braces in (2.8), $-((\partial\varepsilon/\partial\boldsymbol{p})\cdot\nabla(\boldsymbol{p}\cdot\boldsymbol{v}_n))$ we split off the term of the form div $\boldsymbol{v}_n$ which is contained in it and symmetrise the remainder.

The result (2.9) enables us to reach preliminary conclusions about the kinds of kinetic coefficients one might expect in helium II. The terms in (2.9) proportional to div $(\boldsymbol{j}-\varrho\boldsymbol{v}_n)$ and div $\boldsymbol{v}_n$ will correspond to second viscosity; the terms containing ∇T determine a coefficient which is similar to the thermal conductivity and, finally, the terms containing $(\partial v_{nk}/\partial r_i)+(\partial v_{ni}/\partial r_k)-\frac{2}{3}\delta_{ik}(\partial v_{nl}/\partial r_l)$ determine the first viscosity due to normal motion. Let us note the interesting fact that there is no first viscosity connected with super-

* The functions A, B, and C depend, in general, on the momentum p of the elementary excitation. However, in all cases where one has solved the kinetic equation for the rotons, these coefficients turn out to be quantities which are independent of p. This can be explained as follows. When the collision integrals are evaluated for rotons only the roton–roton scattering process turns out to be important and this has a probability which according to ref. 2 is independent of the energy of the colliding rotons. In that case the collision integral $J(n)$ can thus in final reckoning be written in the form const. $(n-n_0)$.

fluid flow and due to the excitations, although logically such an effect could be possible, as the condition for superfluidity consists merely in the requirement that curl $\boldsymbol{v}_s = 0$.

Normal density

We evaluate the total momentum of the relative motion in helium II.

$$\overline{n\boldsymbol{p}} = \int n_0 \boldsymbol{p} \, d\tau_p.$$

We substitute here the distribution function (2.1) expanding it in a power series in the difference $(\boldsymbol{p}\cdot\boldsymbol{v}_n)-(\boldsymbol{p}\cdot\boldsymbol{v}_s)$. The zeroth term in the expansion naturally gives zero when substituted into the integral. The first-order term gives the momentum of the relative motion

$$\overline{n\boldsymbol{p}} = \int n' \frac{1}{kT}[(\boldsymbol{p}\cdot\boldsymbol{v}_s)-(\boldsymbol{p}\cdot\boldsymbol{v}_n)]\boldsymbol{p}\, d\tau_p = \frac{1}{3kT}(\boldsymbol{v}_s-\boldsymbol{v}_n)\int n'p^2\, d\tau_p. \tag{2.10}$$

$\overline{n\boldsymbol{p}}$ is clearly equal to the difference between the total momentum $\boldsymbol{j} = \varrho_n\boldsymbol{v}_n+\varrho_s\boldsymbol{v}_s$ and the momentum of the superfluid motion $\varrho\boldsymbol{v}_s$. Thus

$$\overline{n\boldsymbol{p}} = \boldsymbol{j}-\varrho\boldsymbol{v}_s = \varrho_n(\boldsymbol{v}_n-\boldsymbol{v}_s) = -\frac{1}{3kT}(\boldsymbol{v}_n-\boldsymbol{v}_s)\int n'p^2\, d\tau_p.$$

Hence, we finally find

$$\varrho_n = -\frac{1}{3kT}\int n'p^2\, d\tau_p. \tag{2.11}$$

Similarly we can also obtain the next terms in the expansion of ϱ_n, of order $(\boldsymbol{v}_n-\boldsymbol{v}_s)^2$, and so on.

Entropy of helium

The energy flux in helium II in the approximation which is linear in $\boldsymbol{v}_n$ is in the case when the total momentum $\boldsymbol{j} = 0$ equal to

$$\boldsymbol{Q} = ST\boldsymbol{v}_n. \tag{2.12}$$

The same flux is, according to (1.23), equal to

$$Q = \overline{nH \frac{\partial}{\partial \boldsymbol{p}} H}. \tag{2.13}$$

Comparing (2.12) and (2.13) we find a relation which determines the entropy S of the helium

$$ST\boldsymbol{v}_n = \int nH \frac{\partial}{\partial \boldsymbol{p}} H \, d\tau_p. \tag{2.14}$$

We substitute here the distribution function (2.1) after expanding it in powers of $(\boldsymbol{p}\cdot\boldsymbol{v}_n)/kT$. The zeroth-order term when substituted into the integral gives zero. The linear term gives

$$ST\boldsymbol{v}_n = -\frac{1}{kT}\int n'\left(\frac{H}{kT}\right) H \frac{\partial}{\partial \boldsymbol{p}} H \, d\tau_p (\boldsymbol{p}\cdot\boldsymbol{v}_n).$$

We neglect here all terms of higher than first order in $\boldsymbol{v}_n$ and $\boldsymbol{v}_s$, and we get finally

$$S = -\frac{1}{3kT^2}\int \varepsilon n'\left(\frac{\partial \varepsilon}{\partial \boldsymbol{p}}\cdot\boldsymbol{p}\right) d\tau_p. \tag{2.15}$$

We now elucidate the additional terms in the hydrodynamic equations which are due to the non-equilibrium conditions. We must bear in mind the conditions

$$\int (n-n_0)\boldsymbol{p}\, d\tau_p = 0, \quad \int (n-n_0)\varepsilon\, d\tau_p = 0, \tag{2.16}$$

which follow from the requirement that the non-equilibrium function n must lead to the same values of the total energy and total momentum as the equilibrium function n_0.

In the general case the first of conditions (2.16) imposes a restriction on the coefficient B found when we solved the kinetic equation. It is automatically satisfied if the magnitude of the coefficient B is independent of the momentum p. We then have according to (2.9),

(2.11), and (2.15)

$$\int (n-n_0)\boldsymbol{p}\,d\tau_p = B\int \frac{n'}{kT}\left(\frac{\nabla T}{T}\cdot\left(\boldsymbol{p}\frac{ST}{\varrho_n}-\varepsilon\frac{\partial\varepsilon}{\partial\boldsymbol{p}}\right)\right)\boldsymbol{p}\,d\tau_p$$

$$= B\frac{\nabla T}{3kT^2}\int n'p\left(\frac{ST}{\varrho_n}p-\varepsilon\frac{\partial\varepsilon}{\partial p}\right)d\tau_p$$

$$= B\frac{\nabla T}{3kT^2}\left\{\frac{ST}{\varrho_n}\int n'p^2\,d\tau_p-\int n'\varepsilon\frac{\partial\varepsilon}{\partial p}p\,d\tau_p\right\}=0. \qquad (2.17)$$

Since according to (2.16) the magnitude of the momentum $\overline{pn}$ does not change the additional terms do not occur in the equation of continuity (2.10). In the equations of motion (1.9) we get additional terms in the first approximation:

$$\frac{\partial}{\partial r_k}\left\{\int (n-n_0)p_i\frac{\partial\varepsilon}{\partial p_k}\,d\tau_p+\delta_{ik}\int (n-n_0)\frac{\partial\varepsilon}{\partial p}p\,d\tau_p\right\}. \qquad (2.18)$$

Substituting (2.9) into the integral (2.18) we find after some simple transformations and taking the symmetry properties of these integrals into account

$$\frac{\partial}{\partial r_i}\left\{\operatorname{div}(\boldsymbol{j}-\varrho\boldsymbol{v}_n)\int A\left(\frac{\varepsilon}{T}\frac{\partial T}{\partial\varrho}-\frac{\partial\varepsilon}{\partial\varrho}\varrho\right)\frac{n'}{3kT}\left[\left(\boldsymbol{p}\cdot\frac{\partial\varepsilon}{\partial\boldsymbol{p}}\right)\frac{1}{3}+\frac{\partial\varepsilon}{\partial\varrho}\varrho\right]d\tau_p\right.$$

$$+\operatorname{div}\boldsymbol{v}_n\int A\left(\frac{\varepsilon}{T}\frac{\partial T}{\partial\varrho}\varrho+\frac{\varepsilon}{T}\frac{\partial T}{\partial S}S-\frac{\partial\varepsilon}{\partial\varrho}\varrho\right.$$

$$\left.\left.-\frac{1}{3}\left(\frac{\partial\varepsilon}{\partial\boldsymbol{p}}\cdot\boldsymbol{p}\right)\right)\frac{n'}{3kT}\left[\frac{1}{3}\left(\boldsymbol{p}\cdot\frac{\partial\varepsilon}{\partial\boldsymbol{p}}\right)+\frac{\partial\varepsilon}{\partial\varrho}\varrho\right)d\tau_p\right\}$$

$$+\frac{\partial}{\partial r_k}\frac{1}{2}\left(\frac{\partial v_{ni}}{\partial r_k}+\frac{\partial v_{nk}}{\partial r_i}-\frac{2}{3}\delta_{ik}\frac{\partial v_{nl}}{\partial r_l}\right)\int C\left[\left(\frac{\partial\varepsilon}{\partial\varrho}\right)^2\varrho^2\right.$$

$$\left.-\frac{1}{3}\left(\frac{\partial\varepsilon}{\partial\boldsymbol{p}}\cdot\boldsymbol{p}\right)^2\right]\frac{n'}{5kT}\,d\tau_p. \qquad (2.19)$$

Equation (1.9′) thus becomes

$$\frac{\partial}{\partial t} j_i + \frac{\partial}{\partial r_k}\left\{-\delta_{ik}\zeta_1 \operatorname{div}(\boldsymbol{j}-\varrho \boldsymbol{v}_k)-\delta_{ik}\zeta_2 \operatorname{div} \boldsymbol{v}_n - \eta\left(\frac{\partial v_{ni}}{\partial r_k}+\frac{\partial v_{nk}}{\partial r_i}\right.\right.$$

$$\left.-\frac{2}{3}\delta_{ik}\frac{\partial v_{nl}}{\partial r_l}\right)+\delta_{ik}p_0+\overline{n_0 p_i \frac{\partial \varepsilon}{\partial p_k}}+\delta_{ik}\overline{n_0 \frac{\partial \varepsilon}{\partial \varrho}}\varrho+\overline{n_0 p_i}v_{sk}+\overline{n_0 p_k}v_{si}$$

$$\left.+\varrho v_{sk}v_{si}\right\} = 0. \qquad (2.20)$$

The coefficients ζ_1 and ζ_2 which have the physical meaning of second viscosity coefficients are, according to (2.19), equal to

$$\zeta_1 = -\int A\frac{n'}{kT}\left(\frac{\varepsilon}{T}\frac{\partial T}{\partial \varrho}-\frac{\partial \varepsilon}{\partial \varrho}\right)\left(\frac{\partial \varepsilon}{\partial \varrho}\varrho+\frac{1}{3}\left(\frac{\partial \varepsilon}{\partial \boldsymbol{p}}\cdot\boldsymbol{p}\right)d\tau_p;\right. \quad (2.21)$$

$$\zeta_2 = -\int A\frac{n'}{kT}\left[\frac{\varepsilon}{T}\left(\frac{\partial T}{\partial \varrho}\varrho+\frac{\partial T}{\partial S}S\right)-\frac{\partial \varepsilon}{\partial \varrho}\varrho\right.$$

$$\left.-\frac{1}{3}\left(\frac{\partial \varepsilon}{\partial \boldsymbol{p}}\cdot\boldsymbol{p}\right)\right]\left[\frac{\partial \varepsilon}{\partial \varrho}\varrho-\frac{1}{3}\left(\frac{\partial \varepsilon}{\partial \boldsymbol{p}}\cdot\boldsymbol{p}\right)\right]d\tau_p. \qquad (2.22)$$

The coefficient η is the first viscosity coefficient; it is according to (2.19) equal to

$$\eta = -\frac{1}{2}\int\frac{n'C}{5kT}\left[\left(\frac{\partial \varepsilon}{\partial \boldsymbol{p}}\right)^2 p^2-\frac{1}{3}\left(\frac{\partial \varepsilon}{\partial \boldsymbol{p}}\cdot\boldsymbol{p}\right)^2\right]d\tau_p =$$

$$= -\frac{1}{15kT}\int Cn'\left(\frac{\partial \varepsilon}{\partial \varrho}\right)^2\varrho^2\,d\tau_p. \qquad (2.23)$$

In the non-equilibrium case there appears in the superfluid motion equation (1.15) an additional term

$$\frac{\partial}{\partial r_i}\int(n-n_0)\frac{\partial \varepsilon}{\partial \varrho}\tau_p. \qquad (2.24)$$

Substitution of expression (2.9) for the difference $n-n_0$ into this integral gives similar to (2.19)

$$\frac{\partial}{\partial r_i}\left\{\operatorname{div}(\boldsymbol{j}-\varrho \boldsymbol{v}_n)\int A\frac{n'}{kT}\left(\frac{\varepsilon}{T}\frac{\partial T}{\partial \varrho}-\frac{\partial \varepsilon}{\partial \varrho}\right)\frac{\partial \varepsilon}{\partial \varrho}d\tau_p\right.$$

$$+\operatorname{div}\boldsymbol{v}_n\int A\frac{n'}{kT}\left[\frac{\varepsilon}{T}\left(\frac{\partial T}{\partial \varrho}\varrho+\frac{\partial T}{\partial S}S\right)-\frac{\partial \varepsilon}{\partial \varrho}\varrho-\frac{1}{3}\left(\frac{\partial \varepsilon}{\partial \boldsymbol{p}}\cdot\boldsymbol{p}\right)\right]\frac{\partial \varepsilon}{\partial \varrho}d\tau_p.$$

$$(2.25)$$

The equation of superfluid motion is thus in the case when there is a deviation from equilibrium of the form

$$\frac{\partial \boldsymbol{v}_s}{\partial t}+\nabla\left\{\frac{v_s^2}{2}+\phi_0+n_0\overline{\frac{\partial \varepsilon}{\partial \varrho}}-\zeta_3 \operatorname{div}(\boldsymbol{j}-\varrho \boldsymbol{v}_n)-\zeta_4 \operatorname{div} \boldsymbol{v}_n\right\}=0. \quad (2.26)$$

The coefficients ζ_3 and ζ_4 have the physical meaning of second viscosity coefficients; according to (2.25) they are equal to

$$\zeta_3=-\int A\frac{n'}{kT}\left(\frac{\varepsilon}{T}\frac{\partial T}{\partial \varrho}-\frac{\partial \varepsilon}{\partial \varrho}\right)\frac{\partial \varepsilon}{\partial \varrho}\,d\tau_p; \quad (2.27)$$

$$\zeta_4=-\int A\frac{n'}{kT}\left[\frac{\varepsilon}{T}\left(\frac{\partial T}{\partial \varrho}\varrho+\frac{\partial T}{\partial S}S\right)-\frac{\partial \varepsilon}{\partial \varrho}\varrho-\frac{1}{3}\left(\frac{\partial \varepsilon}{\partial \boldsymbol{p}}\cdot\boldsymbol{p}\right)\right]\frac{\partial \varepsilon}{\partial \varrho}\,d\tau_p. \quad (2.28)$$

The coefficients ζ_1, ζ_2, ζ_3, and ζ_4 are not independent.* Because of the Onsager relations they are connected by one relation: $\zeta_1=\zeta_4$.

The validity of this relation can be checked most simply in the case when A is independent of the momentum. According to (2.21) and (2.22) we have

$$\zeta_1-\zeta_4=-\int A\frac{n'}{kT}\left\{-\frac{S}{T}\frac{\partial T}{\partial S}\varepsilon\frac{\partial \varepsilon}{\partial \varrho}+\frac{1}{3}\frac{\varepsilon}{T}\frac{\partial T}{\partial \varrho}\left(\frac{\partial \varepsilon}{\partial d}\cdot\boldsymbol{p}\right)\right\}d\tau_p. \quad (2.29)$$

According to (2.15) the second integral in (2.29) is equal to $-AS\,\partial T/\partial \varrho$. As a result (2.15) becomes after once integrating by parts

$$S=\frac{1}{3T}\int n\frac{\partial}{\partial p}\,\varepsilon p^3)4\pi\,dp(2\pi\hbar)^{-3}. \quad (2.30)$$

We differentiate this relation on the left and on the right with respect to ϱ at constant temperature. We get

$$\frac{\partial S}{\partial \varrho}=(2\pi\hbar)^{-3}\frac{1}{3T}\int\left[\frac{\partial n}{\partial \varrho}\frac{\partial}{\partial p}(\varepsilon p^3)+n\frac{\partial}{\partial p}\left(\frac{\partial \varepsilon}{\partial \varrho}p^3\right)\right]4\pi\,dp.$$

* The second-viscosity coefficients were determined in ref. 1 in a somewhat different way. One easily establishes the following relations between the coefficients ζ_n, ζ_S, ζ_n^S, and ζ_S^S, defined in ref. 1 and the coefficients ζ_1, ζ_2, ζ_3, and ζ_4:

$$\zeta_n=\zeta_2\zeta_2-\zeta_1\zeta_S, \qquad \zeta_S=\zeta_1\varrho_S.$$

$$\zeta_n^S=\varrho_S(\zeta_4-\zeta_3\varrho_S), \quad \zeta_S^S=\varrho_S^2\zeta_3.$$

Integrating the second term by parts we find

$$\frac{\partial S}{\partial \varrho} = (2\pi)\hbar^{-3}\frac{1}{3T}\int n'\frac{1}{kT}\left[\frac{\partial \varepsilon}{\partial \varrho}\frac{\partial}{\partial p}(\varepsilon p^3) - \frac{\partial \varepsilon}{\partial \varrho}\left(\frac{\partial \varepsilon}{\partial p}p^3\right)\right] 4\pi\, dp$$

$$= \frac{1}{kT^2}\int n'\frac{\partial \varepsilon}{\partial \varrho}\varepsilon\, d\tau_p. \tag{2.31}$$

According to (2.31) the first integral in (2.29) is equal to

$$A\int \frac{n'}{kT}\frac{\partial \varepsilon}{\partial \varrho}\frac{\varepsilon}{T}\frac{\partial T}{\partial S}S\, d\tau_p = -AS\left(\frac{\partial T}{\partial S}\right)_\varrho\left(\frac{\partial S}{\partial \varrho}\right)_T = AS\left(\frac{\partial T}{\partial \varrho}\right)_S$$

The sum of the first and the second integrals in (2.29) thus vanishes and hence*

$$\zeta_1 = \zeta_4. \tag{2.32}$$

We finally elucidate the form of the additional terms in the energy flux (1.23) which arise when there is no equilibrium. First of all we transform the term $\overline{nH(\partial/\partial \boldsymbol{p})H}$ in expression (1.23) for the flux $\boldsymbol{Q}$. We expand the distribution function n in a power series in $(\boldsymbol{p}\cdot\boldsymbol{v}_n)$. The zeroth term vanishes as we have already noted because of the symmetry of the integrand in $\boldsymbol{p}$-space. The first-order term to which we shall restrict ourselves is equal to

$$-\frac{1}{kT}\int n'\left(\frac{H}{kT}\right)(\boldsymbol{p}\cdot\boldsymbol{v}_n)H\frac{\partial}{\partial \boldsymbol{p}}H\, d\tau_p.$$

Moreover, after integrating by parts we get

$$\overline{nH\frac{\partial}{\partial \boldsymbol{p}}H} = \int n\left((H\boldsymbol{v}_n + (\boldsymbol{p}\cdot\boldsymbol{v}_n)\frac{\partial H}{\partial \boldsymbol{p}}\right) d\tau_p. \tag{2.33}$$

In the first approximation which we are looking for, the additional terms in the energy flux are according to (1.23) and (2.33) equal to

$$(\overline{n_0\boldsymbol{p}} + \varrho\boldsymbol{v}_s)\int (n-n_0)\frac{\partial \varepsilon}{\partial \varrho}d\tau_p + \int (n-n_0)\left[H\boldsymbol{v}_n + (\boldsymbol{p}\cdot\boldsymbol{v}_n)\frac{\partial H}{\partial \boldsymbol{p}}\right]d\tau_q.$$

* *(Note by Editor.)* Petru (1968) has obtained this relation in the Bogolubov scheme (1963).

We now substitute expression (2.9) for $n-n_0$; using the symmetry properties of the integrals obtained which are similar to the earlier ones we find

$$(\overline{n_0 \boldsymbol{p}}+\varrho \boldsymbol{v}_s)\left\{\operatorname{div}(\boldsymbol{j}-\varrho \boldsymbol{v}_n)\int A\frac{n'}{kT}\left(\frac{\varepsilon}{T}\frac{\partial T}{\partial \varrho}-\frac{\partial \varepsilon}{\partial \varrho}\right)\frac{\partial \varepsilon}{\partial \varrho}\,d\tau_p\right.$$

$$\left.+\operatorname{div}\boldsymbol{v}_n\int A\frac{n'}{kT}\left(\frac{\varepsilon}{T}\frac{\partial T}{\partial \varrho}\varrho+\frac{\varepsilon}{T}\frac{\partial T}{\partial S}S-\frac{\partial \varepsilon}{\partial \varrho}\varrho-\frac{1}{3}\left(\boldsymbol{p}\cdot\frac{\partial \varepsilon}{\partial \boldsymbol{p}}\right)\right)\frac{\partial \varepsilon}{\partial \varrho}\,d\tau_p\right\}$$

$$+\boldsymbol{v}_n\left\{\operatorname{div}(\boldsymbol{j}-\varrho \boldsymbol{v}_n)\int A\frac{n'}{kT}\left(\frac{\varepsilon}{T}\frac{\partial T}{\partial \varrho}-\frac{\partial \varepsilon}{\partial \varrho}\right)\left(\varepsilon+\frac{1}{3}\left(\boldsymbol{p}\cdot\frac{\partial \varepsilon}{\partial \boldsymbol{p}}\right)\right)d\tau_p\right.$$

$$+\operatorname{div}\boldsymbol{v}_n\int A\frac{n'}{kT}\left(\frac{\varepsilon}{T}\frac{\partial T}{\partial \varrho}\varrho+\frac{\varepsilon}{T}\frac{\partial T}{\partial S}S-\frac{\partial \varepsilon}{\partial \varrho}\varrho\right.$$

$$\left.\left.-\frac{1}{3}\left(\boldsymbol{p}\cdot\frac{\partial \varepsilon}{\partial \boldsymbol{p}}\right)\right)\left(\varepsilon+\frac{1}{3}\left(\boldsymbol{p}\cdot\frac{\partial \varepsilon}{\partial \boldsymbol{p}}\right)\right)d\tau_p\right\}$$

$$+\frac{1}{2}\int(\boldsymbol{p}\cdot\boldsymbol{v}_n)C\frac{n'}{kT}\frac{\partial H}{\partial p}\left(p_k\frac{\partial \varepsilon}{\partial p_i}-\frac{1}{3}\delta_{ik}\left(\boldsymbol{p}\cdot\frac{\partial \varepsilon}{\partial \boldsymbol{p}}\right)\right)\left(\frac{\partial v_{nk}}{\partial r_i}\right.$$

$$\left.+\frac{\partial v_{ni}}{\partial r_k}-\frac{2}{3}\delta_{ik}\frac{\partial v_{nl}}{\partial r_l}\right)+\frac{\nabla T}{3T}\int B\frac{n'}{kT}\varepsilon\left(\frac{\partial \varepsilon}{\partial \boldsymbol{p}}\cdot\left(\boldsymbol{p}\frac{ST}{\varrho_n}+\varepsilon\frac{\partial \varepsilon}{\partial \boldsymbol{p}}\right)\right)d\tau_p. \quad (2.34)$$

We now use the condition $\int(n-n_0)\varepsilon\, d\tau_p = 0$, and the definition of the second viscosity coefficients (equations (2.21), (2.22), (2.27), and (2.28)) and of the first viscosity coefficient (equation (2.23)). We finally obtain an expression for the additional non-equilibrium terms in the ith component of the energy flux in the following form:

$$-(\boldsymbol{j}-\varrho \boldsymbol{v}_n)_l[\zeta_3\operatorname{div}(\boldsymbol{j}-\varrho \boldsymbol{v}_n)+\zeta_4\operatorname{div}\boldsymbol{v}_n]-v_{ni}[\zeta_1\operatorname{div}(\boldsymbol{j}-\varrho \boldsymbol{v}_n)$$

$$+\zeta_2\operatorname{div}\boldsymbol{v}_n]-\eta v_{nk}\left(\frac{\partial v_{ni}}{\partial r_k}+\frac{\partial v_{nk}}{\partial r_i}-\frac{2}{3}\delta_{ik}\frac{\partial v_{nl}}{\partial r_l}\right)-\varkappa\nabla_i T. \quad (2.35)$$

The first three terms in equation (2.35) are caused by the first and second viscosities. The fourth term determines the additional energy flux in helium II which occurs when there is a temperature gradient. The coefficient $\varkappa$ which is analogous to the thermal conductivity coefficient is according to (2.34) equal to*

$$\varkappa = -\frac{1}{3kT^2}\int Bn'\varepsilon\left(\frac{\partial\varepsilon}{\partial \boldsymbol{p}}\cdot\left(\boldsymbol{p}\frac{ST}{\varrho_n}-\varepsilon\frac{\partial\varepsilon}{\partial \boldsymbol{p}}\right)\right)d\tau_p. \tag{2.36}$$

APPENDIX

Exact formulae for the entropy and the normal density of helium II

If helium II is in an equilibrium state we can obtain exact formulae which determine the energy and the density of the normal density as functions of $\boldsymbol{v}_n$ and $\boldsymbol{v}_s$.

The roton and phonon distribution functions in helium II which is moving uniformly with velocities $\boldsymbol{v}_n$ and $\boldsymbol{v}_s$ are, respectively, equal to

$$n_r = \exp\{-[\varepsilon-(\boldsymbol{p}\cdot\boldsymbol{v}_n-\boldsymbol{v}_s)]/kT\},\quad \varepsilon = \Delta+\frac{(p-p_0)^2}{2\mu}, \tag{1}$$

$$n_{ph} = [\exp\{[\varepsilon-(\boldsymbol{p}\cdot\boldsymbol{v}_n-\boldsymbol{v}_s)]/kT\}-1]^{-1},\quad \varepsilon = cp. \tag{2}$$

We can use these functions to calculate the numbers of rotons and phonons per unit volume. For the rotons we have

$$N_r = (2\pi\hbar)^{-3}\int n_r\,d\tau_p = (2\pi\hbar)^{-3}\int \exp[-\varepsilon+(\boldsymbol{p},\boldsymbol{v}_n-\boldsymbol{v}_s)/kT]\,d\tau_p$$
$$= (2\pi\hbar)^{-3}\int e^{-\varepsilon/kT}\frac{kT}{pw}\sinh\frac{pw}{kT}\,d\tau_p = N_{r0}\frac{kT}{p_0w}\sinh\frac{p_0w}{kT}. \tag{3}$$

Here N_{r0} is the number of rotons per unit volume of helium II at rest; $w = |\boldsymbol{v}_n-\boldsymbol{v}_s|$.

Similarly we find for the phonons, integrating the function n_{ph} first over the absolute magnitude p of the momentum and then

* *(Note by Editor.)* In a later paper Khalatnikov (1952; for recent calculations see Khalatnikov and Chernikova, 1966) has given the temperature dependence of η and $\varkappa$.

over its direction:

$$N_{\text{ph}} = (2\pi\hbar)^{-3}\int n_{\text{ph}} d\tau_p = N_{\text{ph}0}\frac{1}{2}\int\frac{d\cos\theta}{[1-(w/c)\cos\theta]^3}$$

$$= N_{\text{ph}0}\frac{1}{[1-(w/c)^2]^2} \tag{4}$$

(N_{pho} is the number of phonons per unit volume in helium II at rest). Using the distribution function (1) to evaluate the momentum of the relative motion due to the rotons, we have

$$\boldsymbol{p}_{\text{rel, r}} = \int \boldsymbol{p} n_{\text{r}}\, d\tau_p (2\pi\hbar)^{-3} = (\boldsymbol{v}_n - \boldsymbol{v}_s)(2\pi\hbar)^{-3}\int e^{-\varepsilon/kT}\frac{p}{w}$$

$$\cdot\left(\frac{kT}{pw}\right)^2\left(\frac{pw}{kT}\cosh\frac{pw}{kT} - \sinh\frac{pw}{kT}\right)d\tau_p$$

$$= (\boldsymbol{v}_n - \boldsymbol{v}_s)\, N_{\text{r}0}\frac{kT}{w^2}\left(\cosh\frac{p_0 w}{kT} - \frac{kT}{p_0 w}\sinh\frac{p_0 w}{kT}\right). \tag{5}$$

The roton part of the normal density $\varrho_{n\text{r}}$ is thus equal to

$$\varrho_{n\text{r}} = N_{\text{r}0}\frac{kT}{w^2}\left(\cosh\frac{p_0 w}{kT} - \frac{kT}{p_0 w}\sinh\frac{p_0 w}{kT}\right). \tag{6}$$

Similar calculations for the phonons give (integrating first over the absolute magnitude and then over the direction of the momentum $\boldsymbol{p}$):

$$\boldsymbol{p}_{\text{rel, ph}} = (2\pi\hbar)^{-3}\int n_{\text{ph}}\boldsymbol{p}\, d\tau_p = (\boldsymbol{v}_n - \boldsymbol{v}_s)\frac{4E_{\text{ph}0}}{3c^2}\frac{1}{[1-(w/c)^2]^3}, \tag{7}$$

$$\varrho_{n\text{ph}} = \frac{4E_{\text{ph}0}}{3c^2}\frac{1}{[1-(w/c)^2]^3}, \quad E_{\text{ph}0} = \frac{4\pi^5(kT)^4}{15\,(2\pi\hbar)^3 c^3}. \tag{8}$$

The roton part S_{r} of the entropy per unit volume of helium II is equal to

$$S_{\text{r}} = -k\int n_{\text{r}}\ln\frac{n_{\text{r}}}{e}\, d\tau_p = \frac{\partial}{\partial T}\left[kT\int n_{\text{r}}\, d\tau_p\right]$$

$$= S_{\text{r}0}\frac{kT}{p_0 w}\sinh\frac{p_0 w}{kT} + N_{\text{r}0}k\left[\frac{kT}{p_0 w}\sinh\frac{p_0 w}{kT} - \cosh\frac{p_0 w}{kT}\right]. \tag{9}$$

Here S_{r0} is the roton part of the entropy per unit volume of helium II at rest.

Finally, we evaluate the phonon part of the entropy S_{ph} per unit volume of helium II. Using the distribution function (2) we find

$$S_{ph} = k \int \{(n_{ph}+1) \ln (n_{ph}+1) - n_{ph} \ln n_{ph}\} \, d\tau_p$$

$$= \frac{4}{3} \frac{E_{ph0}}{T} \frac{1}{2} \int_{-1}^{1} \frac{d \cos \theta}{[1-(w/c) \cos \theta]^3} = \frac{4}{3} \frac{E_{ph0}}{T} \frac{1}{[1-(w/c)^2]^2}. \quad (10)$$

The total normal density is equal to the sum of the roton and phonon parts; from (6) and (8) we have

$$\varrho_n = \varrho_{nr} + \varrho_{nph} = \frac{4E_{ph}}{3c^2} \frac{1}{[1-(w/c)^2]^3} + N_{ph0} \frac{kT}{w^2} \left(\cosh \frac{p_0 w}{kT} - \frac{kT}{p_0 w} \sinh \frac{p_0 w}{kT} \right). \quad (11)$$

Similarly, the entropy S of helium II is equal to the sum of S_r and S_{ph}; from (9) and (10) we get

$$S = S_{r0} \frac{kT}{p_0 w} \sinh \frac{p_0 w}{kT} + N_{r0} k \left[\frac{kT}{p_0 w} \sinh \frac{p_0 w}{kT} - \cosh \frac{p_0 w}{kT} \right] + \frac{4}{3} \frac{E_{ph0}}{T} \frac{1}{[1-(w/c)^2]^2}. \quad (12)$$

REFERENCES

1. I. M. Khalatnikov, *J. Exp. Theoret. Phys. (USSR)* **20,** 243 (1950).
2. L. D. Landau and I. M. Khalatnikov, *J. Exp. Theoret. Phys. (USSR)* **19,** 637, 709 (1948) (*Collected Papers of L. D. Landau,* Pergamon Press 1965, pp. 494 and 511).

T7. ON THE DERIVATION OF A FORMULA FOR THE ENERGY SPECTRUM OF LIQUID HE[4]*

L. P. PITAEVSKII

IT IS well known that Feynman[(1)], using a wave function of a special form, has obtained an expression for the spectrum of the elementary excitations in liquid He^4. A hydrodynamical derivation of this formula is presented below.†

We shall begin with the Hamiltonian for a quantum liquid in the form[(2)]

$$\hat{H} = \frac{1}{2}\int m\hat{v}n\hat{v}\, d\tau + \hat{H}^1[n], \tag{1}$$

where n is the number of atoms per unit volume and $\hat{H}^1[n]$ is the velocity-independent part of the Hamiltonian. We shall assume it to be a function of n. We set $n = \bar{n} + \delta n$ and expand H in terms of second order in δn. The first-order term drops out and we obtain

$$\hat{H} = \hat{H}^1[\bar{n}] + \frac{m\bar{n}}{2}\int \hat{v}^2\, d\tau + \int \varphi(\mathbf{r}, \mathbf{r}')\, \delta n \delta n'\, d\tau\, d\tau', \tag{2}$$

where φ is the second functional derivative of H^1 with respect to n.

* *J. Exp. Theoret. Phys.* (*U.S.S.R.*) **31**, 536 (1956).

† *(Note by Editor.)* This derivation based on the hydrodynamics of the quantum liquid is valid only for small k if the liquid can be treated as a continuous medium. For $k \approx 1/a$ (a is atomic separation) it can be treated as an interpolation between $k \approx 0$ and $k \approx \infty$.

Transforming now to Fourier components

$$\delta n = \sum_k n_k e^{ikr}$$

and taking into account the equation of continuity in the form

$$\delta \dot{n} + \bar{n} \operatorname{div} \hat{\boldsymbol{v}} = 0, \tag{3}$$

as well as the fact that φ depends only upon $|r-r'|$, we obtain

$$\hat{H} = \hat{H}^1[\dot{n}] + \sum_k \left(\frac{m|\dot{n}_k|^2}{2\bar{n}k^2} + \frac{1}{2}\varphi_k |n_k|^2 V \right). \tag{4}$$

This expression has the form of a sum of Hamiltonians of oscillators with frequencies

$$\omega^2(\boldsymbol{k}) = (k^2 \varphi_k \bar{n}/m)V. \tag{5}$$

For the determination of the φ_k we note that the average value of the potential energy of an oscillator in the ground state is equal to $\hbar\omega/4$, whence

$$^1/_2 \varphi_k \overline{|n_k|^2 V} = {}^1/_4 \hbar\omega(\boldsymbol{k}). \tag{6}$$

It is well known, however, that $S(k) = |\bar{n}_k|^2/\bar{n}$ is the Fourier component of the correlation function for the atoms of a liquid, which can be determined from diffraction experiments. Substituting φ_k from (6) into (5), we find for the excitation energy

$$E(\boldsymbol{k}) = \hbar\omega(\boldsymbol{k}) = \hbar^2 k^2/2mS(\boldsymbol{k}),$$

which agrees with Feynman's result.

In conclusion, I would like to express my thanks to L. D. Landau for his advice.

REFERENCES

1. R. P. FEYNMAN, *Phys Rev.* **94**, 262 (1954).
2. L. D. LANDAU, *J. Exp. Theoret. Phys.* (*U.S.S.R.*) **11**, 592 (1941).

Index

O/A INV NO. 84440 OF 4.5.72